Edv

OR26-893

First published in Great Britain 1987 by
Edward Arnold (Publishers) Ltd, 41 Bedford Square, London WC1B 3DQ

Edward Arnold (Australia) Pty Ltd, 80 Waverley Road, Caulfield East,
 Victoria 3145, Australia

Edward Arnold, 3 East Read Street, Baltimore, Maryland 21202, U.S.A.

British Library Cataloguing in Publication Data

Bickerstaff, Gordon F.
 Enzymes in industry and medicine.—(New studies in biology)
 1. Enzymes
 I. Title   II. Series
 547.7'58   QP601

 ISBN 0-7131-2935-2

Text set in 10/11pt Plantin Compugraphic
by Colset Private Limited, Singapore
Printed and bound in Great Britain by Whitstable Litho, Whitstable, Kent

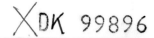

General Preface to the Series

Recent advances in biology have made it increasingly difficult for both students and teachers to keep abreast of all the new developments in so wide-ranging a subject. The New Studies in Biology, originating from an initiative of the Institute of Biology, are published to facilitate resolution of this problem. Each text provides a synthesis of a field and gives the reader an authoritative overview of the subject without unnecessary detail.

The Studies series originated 20 years ago but its vigour has been maintained by the regular production of new editions and the introduction of additional titles as new themes become clearly identified. It is appropriate for the New Studies in their refined format to appear at a time when the public at large has become conscious of the beneficial applications of knowledge from the whole spectrum from molecular to environmental biology. The new series is set to provide as great a boon to the new generation of students as the original series did to their fathers.

1986   Institute of Biology
                                                        20 Queensberry Place
                                                        London SW7 2DZ

Preface

Enzymes are biological catalysts. They have been used for thousands of years to catalyse the various reactions that eventually lead to the production of wine, bread, beer and cheese. The rapid progress of biochemistry over the last 60 years has provided a greater understanding of the role of enzymes in the cell, and has illuminated the fine catalytic properties associated with enzymes. This in turn has led to a substantial increase in the use of enzymes for catalysis in many industrial and medical processes. New developments in genetic engineering, enzyme isolation and enzyme stabilization have combined to increase the range of enzymes available for potential application in biological and chemical reactions.

This book attempts to illustrate some of the traditional and some of the new enzyme applications in analytical procedures, medical therapy and industrial conversions.

I am indebted to Dr John Rudge for supplying the photographic illustration, and to Mr Jim Hamilton and Professor John Smyth for their valuable suggestions and comments. Grateful thanks are also due to Pamela and Natalie for their patience and constant support. The book is dedicated to Emily.

Paisley, 1987   G.F.B.

Contents

General Preface to the Series iii

Preface iv

1 Introduction 1
1.1 Enzymes as catalysts

2 Immobilized Enzymes 7
2.1 Introduction 2.2 Covalent binding 2.3 Adsorption
2.4 Entrapment and encapsulation 2.5 General properties

3 Enzymes as Analytical Tools 20
3.1 Introduction 3.2 Enzyme electrode 3.3 Enzyme analytical reactor 3.4 Diagnostic test strips and dry reagent technology
3.5 Clinical analysis 3.6 Enzyme immunoassay
3.7 Food quality control 3.8 Environmental analysis
3.9 The microbiosensor chip

4 Enzymes in Medical Therapy 46
4.1 Introduction 4.2 Enzyme replacement 4.3 Cancer treatment
4.4 Protein degradation 4.5 Removal of toxic compounds
4.6 Prevention of dental caries

5 Industrial Applications 67
5.1 Introduction 5.2 Sugar production 5.3 Enzymes in detergent products 5.4 Production of semi-synthetic penicillins
5.5 Hydrolysis of lactose 5.6 Production of amino acids
5.7 Cheese manufacture 5.8 Tenderization of meat
5.9 Leather industry 5.10 Fruit juice and wine production
5.11 Baking industry 5.12 Brewing industry
5.13 Recombinant DNA technology

References and Further Reading 94

Index 95

1

Introduction

1.1 Enzymes as catalysts

Enzymes are a specific group of proteins that are synthesized by living cells to function as catalysts for the many thousands of biochemical reactions that constitute the metabolism of a cell. More than 2000 different enzymes are known, and it is likely that many more are awaiting discovery. Enzymes are required in metabolism because at physiological temperature and pH, uncatalysed reactions would proceed at too slow a rate for the vital processes necessary to sustain life. For example, the disintegration of foodstuff by the digestive system involves various enzymes which catalyse the hydrolysis of proteins to amino acids, carbohydrates to sugars, etc., and this is normally accomplished within 3 to 6 hours depending on the amount and type of food. In the absence of enzyme catalysis, such hydrolysis would take 30 years or more to achieve.

In common with all catalysts, enzymes are subject to the normal laws concerning the catalysis of reactions. Thus, the catalyst cannot speed up a reaction that would not occur in its absence, because it is not thermodynamically possible. The catalyst is not consumed during the reaction, and so relatively few catalyst molecules are capable of catalysing the reaction a great many times. Lastly, the catalyst cannot alter the equilibrium position of a given reaction. The vast majority of reactions proceed, eventually, to a state of equilibrium in which the rate of the forward reaction is equal to the rate of the reverse reaction. At equilibrium the substrate and product have specific equilibrium concentrations that are a special characteristic of the reaction. The position of the equilibrium may lie strongly to the product side, for example 1% substrate:99% product, or more toward the substrate side 80%:20%, or near the middle 50%:50%. For example the isomerization of glucose to produce fructose is catalysed by glucose isomerase. Starting from 100% glucose the reaction proceeds to equilibrium, which in this reaction is 45% fructose and 55% glucose. The catalyst cannot change the equilibrium position of the reaction, but it can reduce the time that the reaction normally takes to reach equilibrium. In these respects enzymes are no different from other catalysts. However, enzymes do possess two special attributes that are not found to any great extent in other catalysts, and these are specificity and high catalytic power.

Specificity and the active site

Perhaps the most distinctive feature of enzyme-based catalysis is its specificity. Chemical catalysts display only limited selectivity, whereas enzymes show specificity for the reactants, and also for the susceptible bond involved in the reaction. The degree of specificity for substrate varies from absolute to fairly broad. For example, urease is absolutely specific for its substrate urea (NH_2 $CONH_2$) and structural analogues (e.g. $NH_2 SONH_2$) are not hydrolysed to any great extent by the enzyme. Hexokinase is less specific and shows group specificity for a small set of related sugar molecules.

$$\text{glucose} + \text{ATP} + Mg^{2+} \xrightarrow{\text{hexokinase}} \text{glucose-6-phosphate} + \text{ADP} + Mg^{2+}$$

In addition to glucose, this enzyme will catalyse the phosphorylation of several other sugars such as mannose and fructose, but *not* galactose, xylose, maltose or sucrose. In addition to substrate specificity, enzymes display remarkable product specificity, which ensures that the final product is not contaminated with by-products. Thus, in the above phosphorylation of glucose, the product is exclusively glucose-6-phosphate, and no other phospho-glucose (e.g. glucose-1-phosphate) is produced during the reaction. The formation of by-products by side reactions is a significant problem associated with most of the less specific chemical catalysts. A useful advantage of enzyme specificity is that in a number of cases it extends to selective discrimination between stereo-isomers of a substrate molecule. This stereospecificity is shown by the enzyme D-amino acid oxidase, which is specific for D-amino acids only, and will not catalyse the oxidation of the L-amino acid stereoisomers.

Lastly, some groups of enzymes show a fine specificity for the susceptible bond involved in a reaction. This aspect is neatly illustrated by considering the proteases, which hydrolyse peptide bonds between amino acid residues in poly-peptide chains. Various proteases show specificity for peptide bonds that are located between particular amino acid residues. For example, the blood protease thrombin is very specific, and will only hydrolyse the particular peptide bond between arginine and glycine residues i.e. -arg┤-gly-. The pancreatic protease trypsin will only hydrolyse a peptide bond if it is adjacent to an amino acid residue bearing a positive charge (i.e. lysine or arginine), -lys┤-X- or -arg┤-X-. Other proteases such as pepsin and subtilisin are much less specific, and will catalyse the hydrolysis of a large number of peptide bonds in a polypeptide chain.

Specificity is an inherent feature of enzyme catalysis because the reaction takes place in a particular region of the enzyme that is designed to accommodate the specific participants involved in the reaction. This region is called the active site (see Fig. 1.1), and it is normally a small pocket, cleft or crevice on the surface of the enzyme. It is designed to bring a few of the amino acid residues into contact with the substrate molecule. The site has strong affinity for the substrate because the site amino acid residues are primed for interaction with

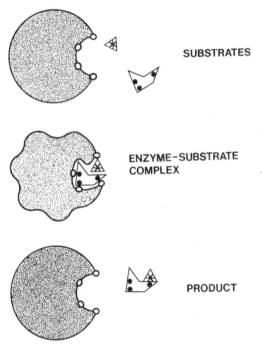

SUBSTRATES

ENZYME-SUBSTRATE
COMPLEX

PRODUCT

Fig. 1.1 Schematic representation of the interaction between an enzyme and its substrate, involving the active site amino acids (**O**).

groups or regions on the substrate molecule. Thus upon substrate binding, an enzyme-substrate (ES) complex is formed, involving non-covalent interactions between the substrate and the site amino acid residues. Consequently the substrate molecule must have the correct shape and/or functional groups to fit into the active site and participate in the interactions. Enzymes with absolute specificity have very precise shape/interaction requirements that are only found in a particular substrate molecule. Enzymes with broad specificity have more flexible active site requirements and can therefore accept a wider range of substrate molecules. The site's amino acid residues participate directly in the catalytic reaction, and are largely responsible for the high catalytic power associated with enzyme reactions (see Fig. 1.1).

Catalytic power

During any reaction the reactants briefly enter a state in which the susceptible substrate bonds are not completely broken and the new bonds in the product are not completely formed (see Fig. 1.2). This transient condition is called the transition state, and it is energy dependent because it requires energy to make and break chemical bonds (about 350 kJ per mole for a covalent bond). This repre-

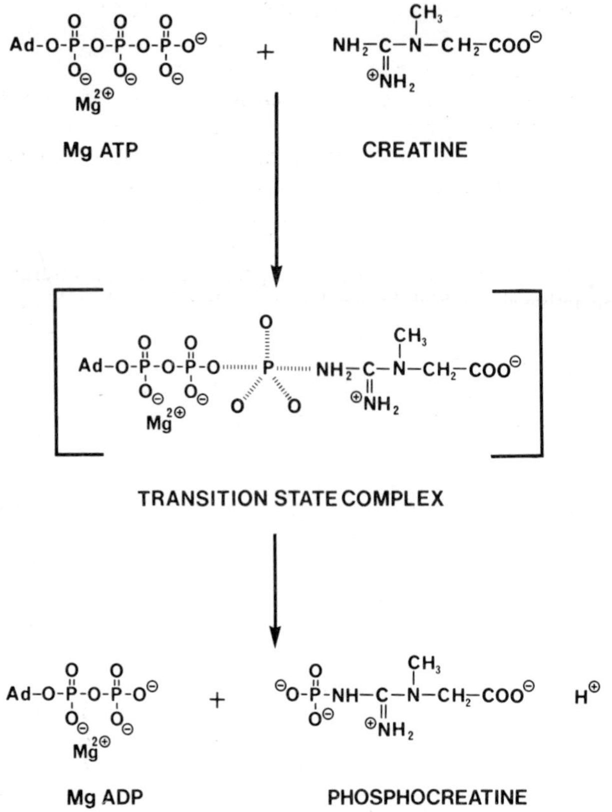

$$\text{Mg ATP} + \text{CREATINE} \xrightarrow{\text{CREATINE KINASE}} \text{Mg ADP} + \text{PHOSPHOCREATINE} + H^+$$

Fig. 1.2 Formation of a transition state complex in the transphosphorylation reaction between ATP and creatine. The reaction requires the participation of magnesium ions.

sents an energy barrier to successful reaction, and is the reason why the vast majority of reactions proceed extremely slowly in the absence of external help. Reactants can be helped towards the transition state by supplying heat energy, high pressure or extreme pH to weaken bonds or by the addition of catalysts. Enzyme catalysts are more effective than chemical catalysts at reducing the energy barrier to facilitate transition state formation and thereby increase the rate of a reaction. The efficiency of enzyme catalysis varies but most enzymes can enhance the rate of an uncatalysed reaction by a factor in the range of 10^5 to 10^{14}. One of the most efficient enzymes is carbonic anhydrase,

$$CO_2 + H_2O \xrightarrow{\text{carbonic anhydrase}} H_2CO_3$$

which catalyses the hydration of up to 600 000 molecules of CO_2 per second under optimal conditions. Carbonic anhydrase is found mostly in red blood cells where it plays a vital role in maintaining the acid-base balance in the body. The enzyme enables rapid transport of molecular CO_2, formed by cellular respiration, from the site of formation (tissues) to the lungs for expulsion. An indication of catalytic power is provided by the turnover number of an enzyme. When an enzyme is fully saturated with its substrate, then the turnover number is the number of substrate molecules converted to product per second. A short list of some turnover numbers is presented in Table 1.1.

The catalytic power of enzymes is due to the precise molecular interactions that occur at the active site, which lower the energy barrier and enable formation of the transition state. There are at least four types of interaction that can accomplish this effect, and they may operate singly or in combination. First, the active site in many enzymes provides a non-polar micro-environment, and the removal of a substrate molecule from an aqueous polar solution into a non-polar phase may alter the conformation of the substrate towards the transition state. Also, a non-polar environment is useful for excluding water molecules, which may interfere in a reaction. Second, the precise alignment of substrate molecules in the active site presents the susceptible bonds at the correct angle so that a collision between reactants will result in the formation of a transition state. Third, the substrate molecule is normally held firmly in the active site by a number of non-covalent interactions, and small movements in the conformation of the enzyme molecule can be transmitted to the active site causing a distortion of the substrate structure, weakening the susceptible bond and reducing the amount of energy required to form a transition state. Lastly, the site amino acid residues contribute catalytic functional groups to participate directly in the reaction. By adding and/or withdrawing electrons, protons (H^+) or other groups, the catalytic functional groups at the active site push the substrate molecules towards the transition state, thereby increasing the rate of reaction. Amino acid side chains such as hydroxyl (–OH), sulphydryl (–SH), carboxyl

Table 1.1 Turnover numbers of some enzymes.

Enzyme	Turnover number (sec^{-1})
Carbonic anhydrase	600 000
Catalase	200 000
Δ^5-3-ketosteroid isomerase	185 000
Acetylcholinesterase	25 000
Beta-amylase	18 000
Penicillinase	2 000
Fumarase	2 000
Lactate dehydrogenase	1 000
Beta-galactosidase	208
Chymotrypsin	100
Phosphoglucomutase	20
Tryptophan synthetase	2
Lysozyme	1

(–COOH) and amino (–NH$_2$) can function as proton donors or acceptors depending on the state of ionization of the side chain. Other side chains exert different chemical influences, and many enzymes employ extra help in the form of a cofactor to supply further chemical influence. For example, in the hydration reaction of CO$_2$ discussed earlier, one atom of zinc per enzyme molecule is essential for enzyme activity.

2
Immobilized Enzymes

2.1 Introduction

Until about 35 years ago, almost all biological catalysis in industrial processes was accomplished using whole cells or tissues, and the great versatility of microbial cells as bio-catalysts has led to the development of the present fermentation industry. During the normal fermentation process micro-organisms grow and multiply, consuming nutrients in the fermentation broth and synthesizing various compounds in addition to the main desired product. Consequently, catalysis by microbial cells is not very efficient, because a high percentage of substrate in the fermentation broth is utilized for growth and the synthesis of various side products. Also the final fermentation liquor is usually a heterogeneous mixture of substances from which the desired product must be isolated.

In recent years there has been an increasing use of isolated enzyme preparations in industrial, analytical and medical procedures. The most obvious advantages are greater efficiency of substrate conversion, higher yields and good product uniformity. However, these advantages must be balanced against the additional costs of enzyme isolation, and the relatively poor stability of purified soluble enzymes. These particular drawbacks have slowed the advancement of enzyme applications, and much research effort has been expended to overcome these problems. Large scale procedures of enzyme isolation have helped to reduce enzyme isolation costs. New procedures of enzyme immobilization have provided useful preparations with greater stability that are also suitable for re-use (see Mosbach, 1976).

In solution, soluble enzyme behaves as any other solute in that it is readily dispersed in the solvent and has complete freedom of movement. Enzyme immobilization may be considered as a technique specifically designed to greatly restrict the freedom of movement of an enzyme. There are four general approaches for achieving enzyme immobilization – adsorption, covalent binding, entrapment and encapsulation – these are illustrated in Fig. 2.1.

Normally the enzyme is attached to or located within an insoluble support material, thus separating the enzyme from the bulk of the solution, and creating a heterogeneous two-phase system as illustrated in Fig. 2.2. Immobilized enzyme technology has received a substantial level of research effort over the last 20 years or so, and a great bank of new information has been accumulated.

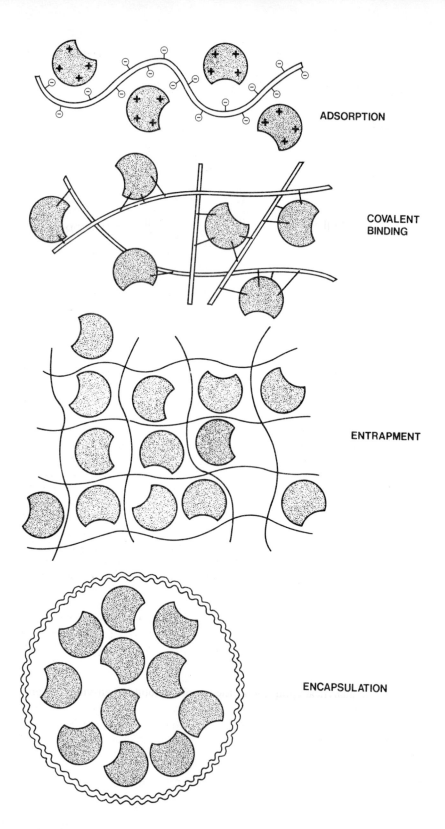

ADSORPTION

COVALENT
BINDING

ENTRAPMENT

ENCAPSULATION

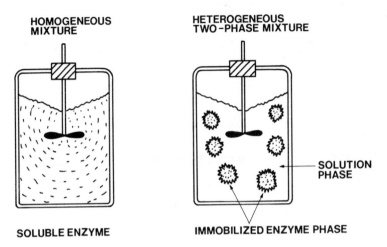

HOMOGENEOUS MIXTURE

HETEROGENEOUS TWO-PHASE MIXTURE

SOLUTION PHASE

SOLUBLE ENZYME

IMMOBILIZED ENZYME PHASE

Fig. 2.2 Diagram showing how immobilization separates an enzyme from the bulk solution phase.

This groundwork has established a firm base for evaluating the usefulness of immobilized enzymes in a wide range of industrial, medical and analytical processes. In addition, the research has identified the distinct advantages that immobilized enzymes have over soluble enzymes. In particular, since an immobilized enzyme is separate from the solution phase, it can be collected after catalysis is completed and re-used again and again. Also the final product is free of enzyme, whereas in soluble enzyme processes the product is contaminated with enzyme, which must be removed to give a pure final product. Immobilization also enables greater control over the catalytic process by removal or addition of immobilized enzyme to vary the rate of reaction. By placing the immobilized enzyme in a cylindrical flow-through device, it is possible to set up a continuous enzyme reactor in which a continuous flow of substrate enters at one end and a continuous flow of product emerges from the other end. Such a system has several economic advantages over the normal batch operation. Lastly, immobilization may improve the stability of an enzyme, and thereby enable use of some enzymes that in soluble form were considered too unstable for practical use.

Immobilized enzyme applications are evolving slowly, particularly in industry, and in many cases this is because the new application normally requires new process equipment. It may not be economically prudent to spend £85 000 on new processing equipment in order to install an immobilized enzyme process that will provide production savings of £5000 per year. However, in medical and analytical applications, the use of immobilized enzymes has progressed rapidly as these areas have been able to make greater use of new developments the costs involved being nowhere near as great as those for large industrial scale applications.

Fig. 2.1 (Left) The principal approaches used for immobilization of enzymes.

2.2   Covalent binding

This method of immobilization involves the formation of a covalent bond between the enzyme and a support material (or matrix). The bond is normally formed between a functional group present on the surface of the matrix, and a functional group belonging to one of the amino acid residues on the surface of the enzyme. A number of amino acid functional groups are suitable for participation in covalent bond formation. Those which are most often involved in covalent binding are: the amino group (NH_2) of lysine or arginine, carboxyl group (CO_2H) of aspartic acid or glutamic acid; hydroxyl group (OH) of serine or threonine; and sulphydryl group (SH) of cysteine.

A large range of support materials is available for covalent binding – some of them are listed in Table 2.1. The extensive range of support materials available reflects the fact that no ideal matrix exists. Therefore, the advantages and disadvantages of a given matrix must be taken into account when considering the appropriate procedure for a given enzyme immobilization. Many factors may influence the selection of a particular matrix, and some of the more important of these are identified below.

(1)   The cost and availability of the matrix can be a limiting factor, particularly in industry, if large quantities of matrix are required.

(2)   The binding capacity of the matrix is the amount of enzyme bound per given weight of matrix, and covalent binding is normally associated with low binding capacity.

(3)   Hydrophilicity is the ability to incorporate water into the matrix, which is important because many enzymes require an aqueous environment for optimum activity and/or stability.

(4)   The complexity of the immobilization procedure may be significant, because some of the methods involve reagents and/or conditions that may be expensive or hazardous.

(5)   The stability of the matrix is often a key factor for consideration. In some applications, chemical stability is essential to resist disruption by reagents and degradation by micro-organisms. In other cases structural rigidity and durability are required to avoid disintegration in process operations.

Research work has shown that hydrophilicity is the most important factor for maintaining enzyme activity in a matrix environment. Consequently polysàc-

Table 2.1   Some support materials for enzyme immobilization.

Inorganic	Organic	Biological
Alumina	Activated clay	Cellulose + derivatives
Nickel oxide	Polystyrene	Dextran + derivatives
Stainless steel	Acrylics	Agarose + derivatives
Glass	Nylon	Starch + derivatives
Silica	Polyacrylamide	Alginate

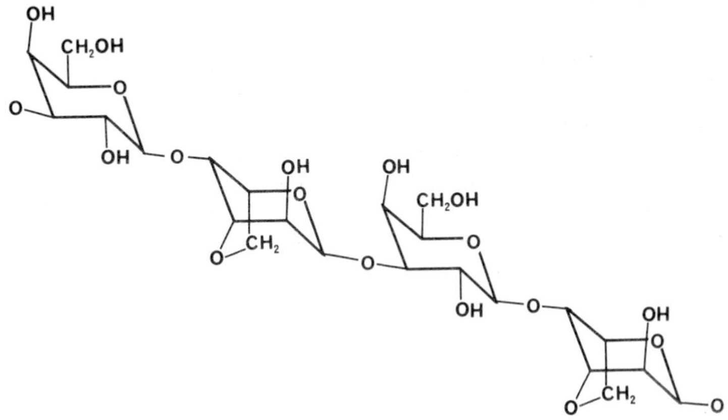

Fig. 2.3 Partial structure of agarose.

charide polymers, which are very hydrophilic, are popular support materials for enzyme immobilization. For example cellulose, dextran (trade name Sephadex), starch, and agarose (trade name Sepharose), are widely used for enzyme immobilization. The sugar residues in these polymers contain hydroxyl groups (see Fig. 2.3), which are ideal functional groups for participating in covalent bonds. Also the hydroxyl groups form hydrogen bonds with water molecules and thereby create an aqueous (hydrophilic) environment in the matrix. In solution the long polysaccharide chains form a three-dimensional network of fibres (Fig. 2.3 and Fig. 2.4 (agarose)), which can trap large quantities of water. The mesh network is produced by cross-linkages between the polysaccharide chains, and the extent of the cross-linking can be varied to produce networks of varying porosity. The polysaccharide supports are susceptible to microbial/fungal disintegration, and organic solvents can cause shrinkage of the gels. The supports are usually used in bead form.

Other popular supports for enzyme immobilization are porous silica and porous glass. The micro-architecture of these is shown in Fig. 2.4. Porous silica consists of small spherical particles of silica fused together in such a way as to form micro-cavities and small channels. The support is normally sold in bead form, and is very strong and durable. Sintered borosilicate glass may be tempered to form a system of uniform channels (see Fig. 2.4). The diameter of the channels depends on the tempering conditions. Porous glass is also durable and resistant to microbial disintegration or solvent distortion. However, these two supports are less hydrophilic than the polysaccharide materials.

There are many reaction procedures for joining an enzyme and a support in a covalent bond. However, most of the reactions fall into the following categories:

(1) formation of an isourea linkage;
(2) formation of a diazo linkage;

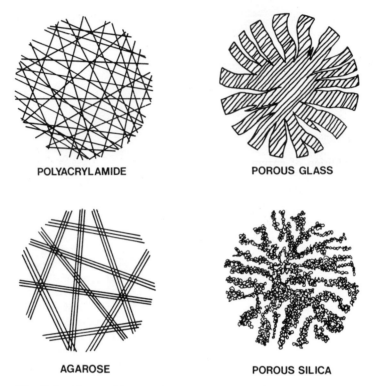

Fig. 2.4   Micro-architecture of some support materials used for enzyme immobilization. (Re-drawn with permission from Koch-Light Laboratories Ltd.)

(3)   formation of a peptide bond; and
(4)   an alkylation reaction.

It is important to choose a method that will not inactivate the enzyme by reacting with amino acids at the active site. So if an enzyme employs a carboxyl group at the active site for participation in catalysis, then it may be wise to choose a reaction that involves amino groups for the covalent bond with the support. Basically two steps are involved in covalent binding of enzymes to support materials. Firstly, functional groups on the support material are activated by a specific reagent, and secondly the enzyme is added in a coupling reaction to form a covalent bond with the support material. Normally the activation reaction is designed to make the functional groups on the support strongly electrophilic (electron deficient), e.g.

$$-\overset{\oplus}{\underset{\underset{O}{\|}}{C}}-O-.$$

In the coupling reaction, these groups will react with strong nucleophiles (electron donating), such as the amino (NH_2) functional groups of certain amino acids on the surface of the enzyme, to form a covalent bond.

An outline of three important methods for covalent immobilization is presented in Fig. 2.5. Cyanogen bromide (CNBr) is aften used to activate the hydroxyl functional groups in polysaccharide support materials. In this method the enzyme and support are joined via an isourea linkage (Fig. 2.5a). In the case of carbodiimide activation (Fig. 2.5b), the support material should have a carboxyl (CO_2H) functional group, and the enzyme and support are joined via a peptide bond. If the support material contains an aromatic amine functional group (Fig. 2.5c), then this can be diazotized using nitrous acid. Subsequent addition of enzyme leads to the formation of a diazo linkage between the reactive diazo group on the support and the ring structure of an aromatic amino acid such as tyrosine.

It is important to realise that no method of immobilization is restricted to a particular type of support material, and that an extremely large number of permutations are possible between methods of immobilization and support materials. This is made possible by the chemical modification of normal functional groups on a support material to produce a range of derivatives containing different functional groups. For example, the normal functional group in cellulose is the hydroxyl group, and chemical modification of this has produced a range of cellulose derivatives such as AE-cellulose (aminoethyl), CM-cellulose (carboxymethyl), DEAE-cellulose (diethylaminoethyl). Thus chemical modification increases the range of immobilization methods that can be used for a given support material. Derivatization can also be used to modify the charges on the surface of a support material, and improve the binding of immobilized enzyme.

2.3 Adsorption

The adsorption of an enzyme onto a support material is the simplest method of obtaining an immobilized enzyme. Basically, it is the adhesion of an enzyme to a support material by non-covalent bonds, and does not require any pre-activation step to alter the functional groups on the support. The bond or bonds formed between an enzyme and support material will be dependent upon the existing surface chemistry of the support. Adsorption normally involves ionic interactions, hydrophobic interactions or hydrogen bonding (see Fig. 2.1).

Many of the support materials available have sufficient surface-charge properties to be suitable for use in immobilization by adsorption. For example alumina, activated carbon, kaolinite, bentonite, porous glass, anion-exchange resins (e.g. DEAE-Sephadex) and cation-exchange resins (e.g. CM-cellulose). Binding of an enzyme to a support is straightforward, and consists of simply mixing an aqueous solution of enzyme with the support material for a period of time after which the excess enzyme is washed away from the immobilized enzyme. The procedure requires strict control of pH and ionic strength because these can alter the charges on the enzyme and the support, and affect the level of adsorption. A simple shift in pH can cancel ionic interactions and release the

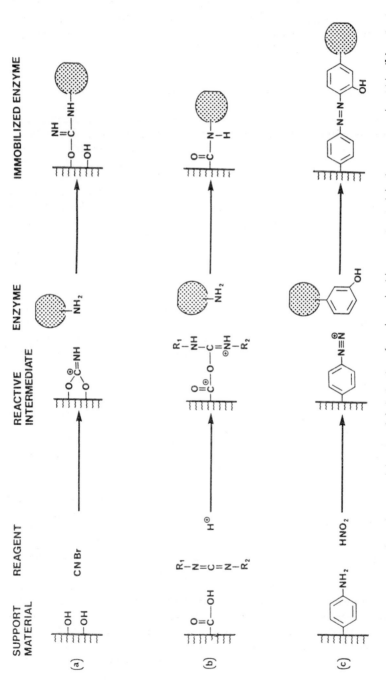

Fig. 2.5 Mechanisms of covalent immobilization. **(a)** Activation of a polysaccharide support material using cyanogen bromide; **(b)** activation of a carboxy derivative support material using carbodiimide; **(c)** activation of a para-aminobenzoyl derivative support material using nitrous acid.

enzyme from the support material. The principal advantages of this method of immobilization are that it is cheap, simple, and enables regeneration of inactive enzyme with fresh enzyme. The main disadvantage is that because the bonding is non-covalent, it is not strong so a certain amount of enzyme leakage occurs from the immobilized enzyme preparation.

2.4 Entrapment and encapsulation

Immobilization by entrapment or encapsulation differs from the other methods of immobilization in that enzyme molecules are free in solution, but restricted in movement by the lattice structure of a gel or by a semi-permeable membrane. (see Fig. 2.1). The porosity of the gel lattice or semi-permeable membrane is controlled to ensure that the structure is tight enough to prevent leakage of enzyme, and at the same time allow free movement of substrate and product.

Entrapment can be achieved by mixing an enzyme with a polymer material and then cross-linking the polymer to form a lattice structure that traps the enzyme. Alternatively, it is possible to mix the enzyme with chemical monomers that are then polymerized to form a cross-linked polymeric network, trapping the enzyme in the interstitial spaces of the lattice. The latter method is more widely used, and a number of acrylic monomers are available for the formation of hydrophilic co-polymers. For example, acrylamide monomer is polymerized to form polyacrylamide and methyl acrylate is polymerized to form polymethacrylate. In addition to the monomer, a cross-linking agent is added during polymerization to form cross-linkages between the polymer chains and help to create a three-dimensional network lattice. The pore size of the gel and its mechanical properties are determined by the relative amounts of monomer and cross-linking agent. It is therefore possible to vary these concentrations to influence the lattice structure. The formed polymer may be broken-up into particles of a desired size or polymerization can be arranged to form beads of defined size.

Encapsulation of enzymes can be achieved by enveloping enzyme molecules within various forms of semi-permeable membranes. Large proteins or enzymes cannot pass out of *or* into the capsule, but small substrates and products can pass freely across the semi-permeable membrane. Many materials have been used to construct micro-capsules varying from 10 μm to 1000 μm in diameter, for example nylon and cellulose nitrate have proved popular. It is also possible to use biological cells as capsules, and a notable example of this is the use of erythrocytes (red blood cells). The membrane of the erythrocyte is normally only permeable to small molecules. However, when erythrocytes are placed in a hypotonic solution they swell-up, stretching the cell membrane and substantially increasing the permeability. In this condition, erythrocyte proteins diffuse out of the cell and enzymes can diffuse into the cell. Returning the swollen erythrocytes to an isotonic solution enables the cell membrane to return to its normal state, and the enzymes trapped inside the cell do not leak out. Another useful method of encapsulation, discussed briefly in Chapter 4 (see Fig. 4.3), is formation of liposomes using lipids to produce a two-layered lipid structure similar to that found in cell membranes. (For practical information on enzyme immobilization see Woodward, 1985 and Trevan, 1980.)

2.5   General properties

The fundamental characteristics of an enzyme catalysed reaction are usually changed in one way or another by immobilization of an enzyme, and the change may be a drawback or an improvement. The nature of the alteration depends on the inherent properties of the enzyme, and additional characteristics imposed by the support material on the enzyme, substrate and product. It is very difficult to quantify these properties and characteristics given the diversity of enzymes, support materials and methods of immobilization. Consequently, it has proved impossible to completely predict what effect a particular immobilization will have on an enzyme or the reaction that it catalyses, and the only recourse is to evaluate a number of methods to discover the system that provides the greatest positive improvement for the application under consideration. The two most important properties that may be changed by immobilization are enzyme stability and the rate of the enzyme catalysed reaction.

Stability is defined as an ability to resist alteration, and in the context of enzyme stability it is important to distinguish several different types of stability. These include resistance to:

inactivation by heat;
disruption by chemicals;
digestion by proteases;
inactivation by change in pH;
loss of activity during storage;
loss of activity due to process operations.

The various types are not necessarily interdependent, and an observed increase in heat stability does not indicate that there will be a corresponding increase in storage stability or operational stability. Although immobilization does not guarantee an improvement in stability, it is generally recognized that it does represent a strategy, which can be used as a means of developing more stable enzyme preparations.

Generally it is found that covalent immobilization is more effective than the other methods of immobilization at improving enzyme resistance to heat, chemical disruption and pH changes. Although the exact details of how these disruptants cause loss of enzyme activity are not fully clear, it is certain that they promote a considerable alteration in the protein structure of an enzyme. In particular, these disruptants probably act by dispersing the many non-covalent bonds responsible for holding the enzyme polypeptide chain in its highly specific conformation or shape, thus causing the polypeptide chain to unfold with a consequent loss of active site structure and catalytic activity (Fig. 2.6). Given that unfolding is necessary for loss of activity to occur, it is probable that multi-point attachment of a polypeptide chain to a support material provides

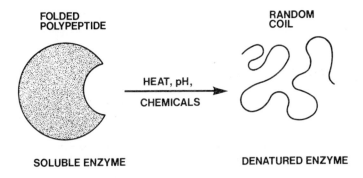

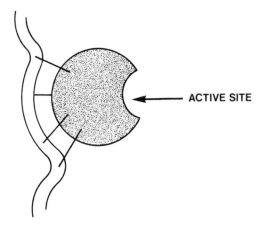

Fig. 2.6 An illustration of the loss of enzyme structure caused by extreme changes of heat and/or pH, or addition of certain chemicals and the possible stabilization of enzyme structure by covalent immobilization.

extra rigidity to the folded protein chain, and therefore greater resistance to unfolding (Fig. 2.6)

Covalent immobilization is also useful for avoiding the natural 'cannibalistic tendencies' exhibited by protease enzymes. A major cause of inactivation with protease enzymes is proteolytic self-degradation (autolysis). This can be avoided by binding protease molecules onto a solid support to prevent them attacking each other. The other methods of immobilization can also contribute to improved enzyme stability, however the relationship between stability and immobilization in these instances is less well understood.

Immobilization almost invariably changes the catalytic activity of an enzyme, and this is clearly reflected in alterations in the characteristic kinetic constants of the enzyme catalysed reaction (see Wilson and Goulding, 1986). In particular the maximum reaction velocity (V_{max}) obtained with an immobilized enzyme is usually lower than that obtained with the corresponding soluble enzyme under the same reaction conditions. The Michaelis constant (Km), which reflects the affinity that the enzyme has for its substrate, is also usually changed upon immobilization, indicating that the active site of the enzyme may be altered, thus affecting the binding of substrate. Four principal factors influencing the catalytic activity of immobilized enzymes have been identified, namely (a) conformation effect; (b) steric effect; (c) micro-environment effect; and (d) diffusion effect.

The conformation of an enzyme refers to the particular shape adopted by the polypeptide chain, which is essential for maintaining the active site structure. Immobilization procedures that involve modification or interaction with amino acid residues on the polypeptide chain can sometimes disturb the structure and thereby affect the enzyme activity. Covalent immobilization is most likely to cause an alteration in the protein conformation of an enzyme. A steric problem arises if the enzyme is immobilized in a position that causes the active site to be less accessible to the substrate molecules. For example, in Fig. 2.6, the active site of the immobilized enzyme is freely accessible. However, if the surface amino acids on the enzyme were such that the enzyme was immobilized with the active site facing into the support material then access to the active site would be restricted and substrate binding would be affected.

In solution, a free enzyme molecule is surrounded by a homogeneous micro-environment in which the enzyme is fully integrated with all components of the solution. Immobilization creates a heterogeneous micro-environment consisting of two phases, i.e. the immobilized enzyme and the bulk of the solution from which the immobilized enzyme is separated. Therefore, all components of the reaction, substrates, products, activators, ions etc., are partitioned between the immobilized enzyme phase and the bulk solution phase. This problem can significantly alter the characteristics of an enzyme reaction even if the enzyme molecule itself is not changed by immobilization. The support material may influence the partitioning effect. For example, CM-cellulose is negatively charged and therefore, attracts hydrogen ions from the bulk solution to the immobilized phase. The increase in hydrogen ion concentration lowers the pH of the immobilized phase and this may alter the enzyme activity. If the support material attracts the substrate then this can improve the catalytic activity. Lastly, reaction rate is also reduced by diffusion restriction. As the substrate is consumed, more substrate must diffuse into the enzyme from the bulk solution. This is normally a problem for all forms of immobilized enzymes, but particularly so for encapsulated enzymes.

Diffusional limitations may be divided into two types, external diffusion restriction and internal diffusion restriction. The external type refers to a zone or barrier that surrounds the support material, called the Nernst layer. Substrate molecules can diffuse into this layer by normal convection and by a passive molecular diffusion. If substrate molecules pass through this layer slowly, this may limit the rate of enzyme reaction. External diffusion restriction can be improved by speeding up the flow of solvent over and through the immobilized enzyme by increasing the stirring rate. Internal diffusion restrictions are due to a diffusion limitation inside the immobilized enzyme preparation. In this case diffusion of substrate molecules occurs by a passive molecular mechanism only, which may be more difficult to overcome if it is a seriously limiting factor. The overall rate of diffusion is markedly influenced by the method of immobilization. Covalent and adsorption procedures cause less diffusion limitation than do entrapment and encapsulation procedures.

3

Enzymes as Analytical Tools

3.1 Introduction

Analysis of biological fluids involves the detection and measurement of many different types of compound, and normally the substance to be analysed (analyte) is a constituent of a complex mixture such as blood serum or the broth from a fermentation. Successful analytical procedures must therefore be specific in order to avoid inaccuracy arising from the presence of other components in the mixture. Also procedures must be sufficiently sensitive to enable detection and measurement of small amounts of analyte. Enzymes are highly specific for their substrates and can catalyse reactions at very low levels of substrate concentration. These inherent characteristics make enzymes very attractive as tools for biochemical analyses in industry and medicine. In view of the large number of enzymes available, and the wide range of reactions catalysed, it is not normally difficult to find an enzyme that will accept a given analyte as a substrate, and provide the basis for an analytical procedure. For example, the enzyme urease is used to estimate the concentration of urea in urine samples.

$$NH_2-\overset{\overset{\displaystyle O}{\|}}{C}-NH_2 + H_2O \xrightarrow{\text{urease}} 2NH_3 + CO_2$$
$$\text{(substrates)} \qquad\qquad\qquad \text{(products)}$$

The concentration of ammonium ions is easily measured, and is related to the concentration of urea.

Although many of the current enzyme-based analytical techniques use soluble enzymes, the trend is moving more towards the use of immobilized enzymes (see Carr and Bowers, 1980 and Hartley *et al.*, 1983). As indicated in Chapter 2, immobilized enzymes tend to have greater resistance to loss of activity. They are also suitable for incorporation into instruments designed for a continuous re-use operation. In particular, immobilized enzymes have featured in the development of several analytical biosensor devices, namely the enzyme electrode, the enzyme analytical reactor, and the diagnostic reagent strip.

3.2 Enzyme electrode

The enzyme electrode may be envisaged as a self-contained analytical biosensor device involving the combination of a thin layer of enzyme with a suitable electrochemical electrode sensor. In practice, the analyte diffuses into the enzyme layer where the catalytic reaction occurs, and the products are formed. The sensor functions as a transducer in that it detects the biochemical change, for example appearance of product, and gives rise to an electrical signal that can be registered by an appropriate meter.

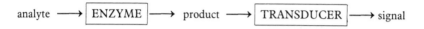

analyte ⟶ ENZYME ⟶ product ⟶ TRANSDUCER ⟶ signal

The signal generated by an electrochemical transducer may be derived from a sensor operating on an amperometric or a potentiometric principle. An amperometric sensor gives rise to a flow of current (micro-amps) between the sensor and a reference electrode, and the current is a linear function of the concentration of analyte. The most notable sensor employing the amperometric principle is the oxygen electrode, which can be used to monitor reactions involving change in dissolved oxygen level. A potentiometric transducer gives rise to a potential difference (milli-volts) between the sensor and a reference electrode. The voltage is a logarithmic function of the concentration of analyte. Both ion-selective electrodes, which detect ionic species (e.g. H^+ and NH_4^+), and gas (e.g. CO_2 and NH_3) sensing electrodes make use of the potentiometric principle.

The fundamental basis for the application of enzyme electrodes is simple to understand. The analyte to be measured is identified (e.g. urea), and an enzyme (e.g. urease), that specifically accepts the analyte as a substrate, is chosen. Knowledge of the enzyme reaction enables the selection of a suitable electrochemical transducer that will detect either the appearance of one of the products of the reaction (NH_4^+ in this case), or the disappearance of one of the substrates. A typical enzyme electrode biosensor device is depicted in Fig. 3.1.

The enzyme layer must be positioned as close as possible to the electrochemical electrode in order to maximize the correspondence between the reaction progress and the electrode response. In simple devices, the enzyme is easily retained in close contact with the active element of the sensing electrode by means of a porous membrane support material, such as cellophane. The support membrane may be held in position against the wall of the electrode with the aid of a rubber 'O'-ring (Fig. 3.1). Alternatively, enzyme electrodes have been constructed with the enzyme immobilized onto the membrane partition of the electrode, and in some cases directly onto the active sensing element of the electrode.

The rapid growth in the development of new enzyme electrode applications is related to advances in immobilized enzyme technology, and the availability of a variety of electrochemical transducers. As more enzymes become commercially available through developments in genetic engineering and enzyme

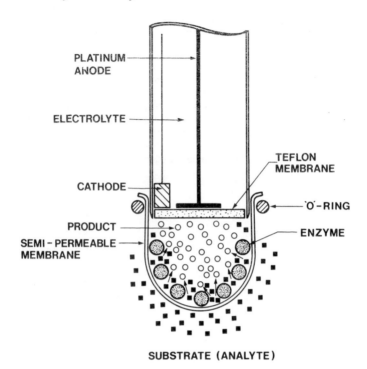

Fig. 3.1 A diagram of a simple enzyme electrode biosensor device, combining an electrochemical electrode and an enzyme immobilized onto a semi-permeable membrane.

purification, it is likely that the growth of enzyme electrode biosensor applications will continue in order to meet the increasingly sophisticated needs of the modern analytical laboratory. A small sample of the range of enzyme electrodes that are available is given in Table 3.1. Most of the simple enzyme electrodes provide a suitable response in the range 10^{-1} to 10^{-4} M analyte concentration, which is satisfactory for most applications. The response time for a measurement depends on the activity of the enzyme and the diffusion rate of analyte into the enzyme layer. Response times vary from 15 seconds to three minutes. An important advantage of the enzyme electrode is the capacity for re-use. In most cases electrodes can be used hundreds or even thousands of times before replacement is necessary. A few of the enzyme electrodes that are used regularly in analysis are discussed more specifically under the various applications later in the chapter.

3.3 Enzyme analytical reactor

In a busy analytical laboratory where many hundreds of analyses are performed every day, there is a strong case for greater automation, particularly for handling large numbers of routine analyses. A variety of automatic analysers

Table 3.1   Some enzymes used in enzyme electrodes.

Analyte	Enzyme(s)
Adenosine phosphate	AMP deaminase
Alcohol	Alcohol dehydrogenase
L-asparagine	L-asparaginase
Cephalosporin	Cephalosporinase
Cholesterol	Cholesterol esterase, cholesterol oxidase
Galactose	Galactose dehydrogenase
Glucose	Glucose oxidase, peroxidase
Glucose	Glucose dehydrogenase
Lactose	Beta-galactosidase, galactose dehydrogenase
Penicillin	Penicillinase
Phenol	Phenol hydroxylase
Phospholipid	Phospholipase, choline oxidase
Sucrose	Invertase, mutarotase, glucose oxidase
Urea	Urease

are available that employ enzymes, and such devices are now standard equipment in most clinical biochemistry laboratories. However, many of the current devices use soluble enzymes together with colour producing reagents. The enzyme and the colour-producing reagent are not re-usable, and as such are consumables that increase the running costs of these devices. A potential approach for reducing the enzyme costs in such systems would be the replacement of the soluble enzyme with an immobilized enzyme to facilitate re-use.

Most of the current range of automatic analysers operate on a continuous flow of samples and reagents through a system of channels to mixing chambers, then on to detecting devices for estimation of the reaction. A typical system is outlined in Fig. 3.2. The most appropriate format of immobilized enzyme for such systems is the enzyme reactor. This is a self-contained flow-through compartment with the enzyme retained by immobilization. The substrate samples flow into the compartment, reaction occurs, and the products flow out of the opposite end of the compartment. In practical terms the enzyme reactor would replace the mixing chamber in the conventional automatic analyser. Two types of enzyme reactor have been investigated for use in continuous flow analysis – the tubular enzyme (coil) reactor and the packed bed (column) reactor (Fig. 3.3).

In the tubular system, the enzyme is attached onto the inner surface of a length of tubing. (Nylon tubing is very suitable.) In the packed bed reactor the enzyme is immobilized, for example onto porous beads or within fibres, and then the immobilized enzyme is packed into a glass or plastic tube to form a small enzyme reactor column. Both systems are effective in replacing the soluble enzyme, and both systems have their own merits and disadvantages. These are taken into consideration when evaluating the performance of enzyme or the analytical device as a whole. The performance characteristics, for example response time, are broadly similar to those given for the enzyme electrode (p. 22).

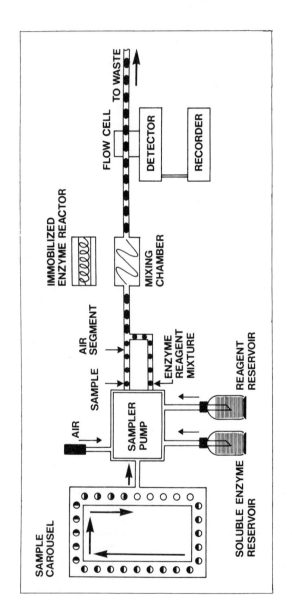

Fig. 3.2 Outline of a simple flow system for automated analysis.

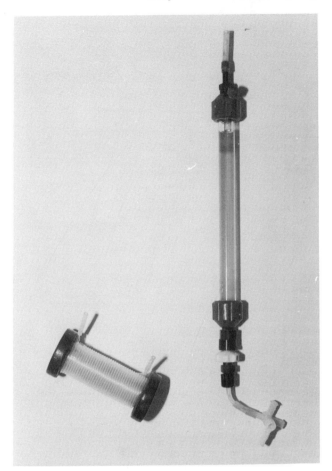

Fig. 3.3 A column (packed bed) reactor (*right*) containing alcohol dehydrogenase covalently bound to agarose, and an enzyme coil (tubular) reactor (*left*) containing glucose dehydrogenase bound to nylon tubing. The reactors were used in the analysis of ethanol and glucose respectively.

3.4   Diagnostic test strips and dry reagent technology

Over the past decade or so, immobilized enzymes have been used in the development of an exciting innovation in clinical biochemistry. The diagnostic test strip is simple to use, inexpensive, small, and capable of fast analysis in the doctor's surgery, or in the home. Basically, the device consists of a thin strip of plastic (8 cm × 1 cm) with a small pad, containing the enzymes and colour reagents, located at one end of the strip (Fig. 3.4). The first of these diagnostic

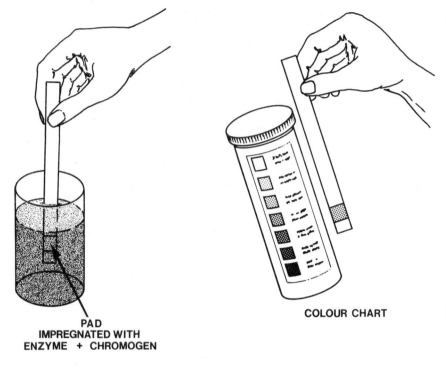

PAD
IMPREGNATED WITH
ENZYME + CHROMOGEN

COLOUR CHART

Fig. 3.4   Application of a diagnostic test strip reagent for semi-quantitative analysis.

strips was a test for glucose in urine. The pad responded with a colour change within 15 seconds after dipping into a sample of urine containing glucose. The colour produced in the pad can be matched against a graded colour chart, supplied with the strips, to give a semi-quantitative estimate of the concentration of glucose in urine. In this case, the pad contains two enzymes (glucose oxidase and peroxidase), immobilized onto a cellulose mat together with the appropriate colour generating dye. The pads are coated with an ethyl cellulose membrane, which is readily permeable to small molecules such as glucose. This coating prevents large structures such as proteins or cells from binding to the cellulose pad and causing interference.

A range of dip-and-read diagnostic test strips are available for rapid testing of blood and urine. Many of the test strips on the market have four or five separate pads on the one strip. This enables the simultaneous estimation of several variables such as glucose, total protein, urea, pH and cholesterol. These test strips have an important use in the doctor's surgery for supporting the preliminary diagnosis of disease and disorder. The test strips for glucose have provided diabetic patients with the means of enabling greater management and control of diabetes.

The use of dry reagents has been further developed by the Eastman Kodak

Corporation to produce a range of multi-layered dry reagent slides containing all the necessary reagents for a given analysis. The production technology is derived from the Corporation's experience in the construction of photographic colour film, which requires up to 15 thin layers of chemicals reproducibly coated onto a transparent support material. An illustration of the structure of a dry reagent slide is given in Fig. 3.5a. The construction varies depending on the nature of the analyte to be measured and the reaction employed in the analysis. The principles involved are best outlined by considering a specific example such as the dry reagent slide for urea analysis.

The urease slide consists of five thin layers sandwiched between two supportive slide mounts. The base layer is a transparent plastic such as polyethylene terephthalate, and above the base layer is an indicator layer of cellulose acetate containing immobilized leuco dye reagent. Above the indicator layer is a semipermeable membrane layer composed of cellulose acetate butyrate, which prevents contaminants in the sample from interfering with the dye reagent. Above this layer is the enzyme reagent layer consisting of urease immobilized onto gelatin together with buffer materials which will provide a pH of 8.0 when water in the sample forms a liquid layer on the gelatin. On top of the reagent layer there is a spreading/reflecting layer composed of porous cellulose acetate containing bound titanium dioxide for reflecting light.

(a)

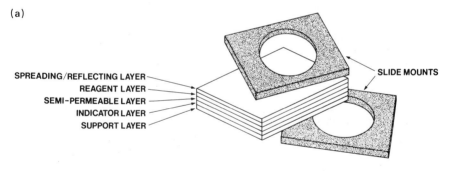

SIZE APPROX. 2·8 x 2·3 cm. x 0·1cm.

(b)

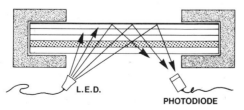

Fig. 3.5 **(a)** Multi-layered dry reagent slide. **(b)** Basic principle involved in reflectance densitometry.

In principle, when a small drop (ten microlitres) of undiluted blood is placed on the slide, it spreads uniformly and rapidly by capillary attraction. The liquid penetrates down to the reagent layer where immobilized urease hydrolyses urea to ammonia and bicarbonate. The ammonia diffuses through the semi-permeable membrane and reacts with the dye reagent to produce a coloured product. The analysis is conducted in a special instrument that is equipped to measure the colour intensity by reflectance densitometry through the transparent base. The instrument employs a light emitting diode as a source and a photodiode as a detector (Fig. 3.5b). The system is simple to operate as slides are disposable, no reagent or buffer solutions require mixing, and the instrument has built-in microprocessors to control the instrument functions, compute the results and print out results.

A few teething problems have arisen such as the need to dispense ten microlitres accurately and with reproducibility. The water volume must be kept low to avoid over-swelling and distortion of the fine layers. Also, patients with higher than normal concentration of blood protein have a more viscous blood that does not spread out evenly over the slide. However, it is likely that this simple and normally precise system will be of use for emergency 'on-call' paramedical organizations attending accidents etc., as well as in the doctor's surgery/general clinics where blood samples are regularly taken for analytical purposes.

3.5   Clinical analysis

The clinical biochemistry laboratory is a key component of the modern hospital, and its principal function is to support the work of the clinician by providing analyses of substances present in biological samples such as blood plasma, urine, faeces, cerebrospinal fluid (CSF), semen and amniotic fluid. Analyses of specimens obtained from patients can contribute information which facilitates the detection and diagnosis of disease or disorder in the body. Analytical data are also of value in the management of diseases since they may contribute to the task of monitoring the progress of medical treatment. For example, it may be vital to determine the concentration of a drug in blood plasma in order to establish the optimum dosage for successful treatment, or provide data on the relationship between drug level and harmful side effects of the drug. In other circumstances, analysis may be concerned with monitoring the level of toxic material arising from a disease, or the disappearance of excess body metabolites as a result of the successful treatment of a disorder.

A large number of different types of analyses are performed, and the analytical laboratory associated with a large general hospital can normally expect to perform at least 20 000 analyses each week. Samples for analysis may be derived from both hospitalized patients and out-patients of the hospital or local health centres. As much as 20% of these analyses may involve the participation of an enzyme as a reagent. Therefore, enzymes have a significant role in clinical analyses, and the current trend is to make further use of enzymes in analytical procedures. A few examples of current enzyme-based analytical techniques will be discussed here in order to illustrate the principles employed in the development and operation of such systems.

Blood glucose

The estimation of glucose in blood serum is one of the most frequent of the many analyses that are undertaken by analytical laboratories. In a normal week, a laboratory may be required to perform some 1500 blood glucose estimations.

Glucose arises in the body largely from the digestion of carbohydrate food-stuffs in the diet. It is oxidized by cells in the body, producing chemical energy (ATP) to drive cellular activities. A stable blood glucose concentration of between 80 and 100 mg per 100 cm^3 is essential to support glucose metabolism and provide a steady energy supply for the body. A fairly constant level of blood glucose is maintained by a set of sophisticated mechanisms involving several hormones, including insulin. If the blood glucose level falls below 50 mg per 100 cm^3 (hypoglycaemia), then the brain signals distress in the form of cold sweats, nausea and loss of concentration. If the level of glucose in the blood rises above 180 mg per 100 cm^3 (hyperglycaemia), then the blood becomes too acidic, and glucose is excreted in the urine. In both cases, if the condition is not corrected the consequences can be fatal.

Insulin plays an important role in the regulation of blood glucose concentration, and in the disease diabetes mellitus the production of insulin ceases or is substantially reduced. When this occurs regulation is impaired because, in the absence of insulin, cells and tissues are unable to take-up glucose from the blood, so blood glucose level rises. At first diagnosis a diabetic may have a blood glucose level as high as 700 mg per 100 cm^3.

In a normal person, blood glucose is automatically monitored by the pancreas, and as the glucose level rises (e.g. after a meal), the pancreas automatically releases the correct amount of insulin into the blood. The insulin promotes glucose take-up by cells, lowering the blood glucose level to the normal range. In diabetes it is necessary to monitor the blood glucose level

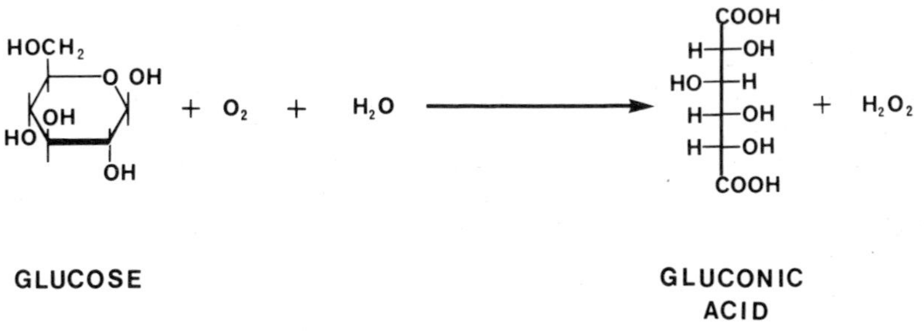

Fig. 3.6 Reaction catalysed by glucose oxidase. The enzyme catalyses the oxidation of glucose to gluconolactone (not shown) which hydrolyses spontaneously to form gluconic acid. Oxygen is utilized as a hydrogen acceptor to form hydrogen peroxide.

manually, and administer the correct amount of insulin by injection. Diabetes mellitus is a comparatively common disease with almost two per cent of the population of Britain having this or a related disorder in the regulation of blood glucose level. Consequently, there is a very large demand for the analysis of glucose in blood and in urine.

The very early methods of blood glucose analysis were chemical methods based on the reducing property of glucose. However, the chemical reagents were not specific for glucose because other components in the blood were capable of reducing the chemical chromogen. The subsequent application of the enzyme glucose oxidase did not immediately overcome the lack of specificity. The enzyme is specific enough for glucose (Fig. 3.6), but the methods that were employed for generating a signal from the product (H_2O_2) were not very specific for H_2O_2. Therefore, it became apparent that it would be necessary to use an additional enzyme to ensure the specificity of the signal generation. The enzyme peroxidase provided the further specificity.

$$H_2O_2 + \text{chromogen} \xrightarrow{\quad \text{peroxidase} \quad} \text{oxidized chromogen} + H_2O$$
$$\text{(colourless)} \qquad\qquad\qquad\qquad \text{(blue)}$$

The necessity of a coupled enzyme system involving two or more enzymes can weigh against the general advantages of using enzymes, particularly if any of the enzymes are expensive. In this case, the enzymes are readily available in a pure form at reasonable cost. (For further discussion on coupled enzyme biosensors see Renneberg *et al.*, 1986.)

An interesting advance has been made in the enzymic analysis of glucose, with the discovery of a bacterial enzyme called glucose dehydrogenase. This enzyme combines specific detection and specific signal generation.

$$\text{glucose} + NAD^+ \xrightarrow[\quad\text{dehydrogenase}\quad]{\text{glucose}} \text{gluconic acid} + NADH + H^+$$

The enzyme requires the participation of a special substrate molecule called NAD^+ (nicotinamide adenine dinucleotide) to act as a hydrogen acceptor. The reduced molecule (NADH) also functions as a signal generator, because it can be readily detected using a spectrophotometer at 340 nm. It is likely that this enzyme will gradually replace the present coupled enzyme systems for the more quantitative analysis of glucose. However, the coupled enzymes/chromogen system may continue to have extensive application in the semi-quantitative diagnostic test strips.

Blood cholesterol

About half of all deaths in the developed countries are caused by artherosclerosis (hardening of the arteries) (see Brown and Goldstein, 1984). In this disease fatty deposits, containing mostly cholesterol, accumulate on the walls of arteries. The deposits harden and enlarge to form plaques (raised circular structures), gradually narrowing the bore of the artery, and restricting the flow of blood. The irregular and rough textured plaques represent a form of tissue damage, and as such cause formation of blood clots (thrombosis). Eventually a blood clot forms in one of the arteries serving the heart, thus starving the heart of oxygen and causing a heart attack.

The link between blood cholesterol and atherosclerosis is now well established. If a person's blood cholesterol level is normally below 150 mg per 100 cm³, then that person should be fairly safe from the possibility of a heart attack. If blood cholesterol level is regularly above 250 mg per 100 cm³, then the risk of serious heart disease is substantially increased. The blood cholesterol level of the average British male is around 220–260 mg per 100 cm³. In many cases heart disease can be controlled or even cured with effective therapy involving regular monitoring of blood cholesterol level.

Analysis of blood cholesterol is also important in the diagnosis and monitoring of several other diseases, in particular, familial hypercholesterolaemia. Patients with this disease have very high levels of plasma cholesterol (650 to 1000 mg per 100 cm³), and the cholesterol is deposited in skin and various other areas of the body. Patients develop artherosclerosis by the early teens, and normally die from a heart attack before the age of 20. The disease is caused by the absence of cell receptor proteins responsible for promoting uptake of cholesterol from the blood into the body tissues (see Brown and Goldstein, 1986).

Quantitative determination of serum cholesterol can be made by a colorimetric test. The procedure is based on the reaction of acetic anhydride with cholesterol in a chloroform solution to produce a characteristic blue-green colour. The reaction is not specific for cholesterol and proteins such as haemoglobin cause interference. It is therefore necessary to extract the cholesterol with an alcohol-ether mixture, which dissolves the cholesterol and precipitates

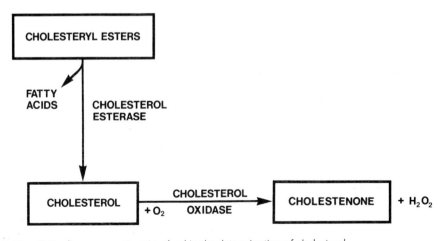

Fig. 3.7 Enzyme reactions involved in the determination of cholesterol.

the protein. The organic solvent must then be removed by evaporation, and the cholesterol taken up in chloroform before the test can be performed.

An enzyme reactor device has been developed that can be used directly with blood serum and therefore obviates the extraction procedure. Cholesterol occurs in blood serum as free cholesterol and as cholesteryl esters. The bio-sensor determines the total cholesterol by employing two enzymes. Cholesterol esterase hydrolyses the esters to free cholesterol, and cholesterol oxidase detects total free cholesterol. Cholesterol oxidase converts cholesterol to cholestenone and H_2O_2, which can be measured amperometrically (Fig. 3.7). The system consists of the two enzymes immobilized in an enzyme reactor, and the elec-trodes are positioned close to the reactor. With a response time of two minutes this procedure provides a substantial saving in time compared with the chemical procedure. The estimation of *total* blood cholesterol is made possible by employing a coupled enzyme system.

Pancreatic disease

The pancreas produces and secretes key hormones for metabolic regulation such as insulin, and a pancreatic juice for use in the digestion of food (Fig. 3.8). In particular the juice contains a number of degradative enzymes including several inactive zymogen precursors of proteases (e.g. trypsinogen) that become

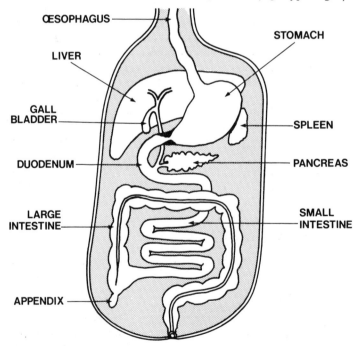

Fig. 3.8 Diagram showing the principal organs associated with digestion.

activated when the juice is released into the duodenum via the pancreatic duct.

Inflammation of the pancreas (pancreatitis) can occur in an acute form or in a chronic form. Acute pancreatitis develops rapidly with severe pain in the middle or left upper abdomen, nausea, vomiting and high temperature, but lasting for a short time. In chronic pancreatitis the symptoms are similar, but the attacks are recurrent and regular. Pancreatitis is caused by premature activation of the pancreatic enzymes producing digestion of the pancreas tissue. This self-digestion process is known to be triggered by various primary agents such as disease of the gall bladder, excessive consumption of drugs (for instance alcohol or morphine), and exposure to specific toxic drugs. With chronic pancreatitis there is a gradual destruction of the pancreas tissue leading to the onset of diabetes and other diseases due to dietary insufficiency, and eventually to complete loss of pancreatic function. Therefore early diagnosis is vital if corrective medicine is to save the pancreas from self-digestion.

Diagnosis is complicated by the fact that 'abdominal pain' may arise from various other organs situated near the pancreas (Fig. 3.8), for example duodenal ulcer, gastric ulcer, gallstones, and diseases of the liver. Consequently, diagnosis of pancreatic disease normally requires several clinical investigations, including minor exploratory surgery. Enzymes play an important role in helping the clinician to arrive at a diagnosis. In this case the analysis is concerned with detecting the presence of pancreatic enzyme itself, rather than using an enzyme to detect an analyte compound.

In acute pancreatitis, pancreatic enzymes are released into the bloodstream, and detection of these enzymes provides the basis for a diagnostic test. Two pancreatic enzymes, lipase and alpha-amylase can be readily assayed in blood serum, and elevated levels of these enzymes strongly indicate acute pancreatitis. This analysis is not applicable for chronic pancreatitis as progressive destruction of the pancreas inevitably leads to greatly reduced enzyme production, and consequently less of the enzymes appear in the blood. An alternative procedure, more applicable for the chronic disease, is to detect low levels of pancreatic enzymes in the undigested food that has passed through the intestine. Chymotrypsin is one of the pancreatic proteases, and the detection of lower than normal levels of this enzyme in faeces is a useful aid in the diagnosis of the chronic disease. Also, by monitoring the level of this enzyme in the faeces it is possible to follow the progress of treatment to restore the normal pancreatic functions.

3.6 Enzyme immunoassay

Enzymes are suitable for the detection of small molecules such as sugars, amino acids and steroids because enzymes can accept such molecules as participants in a catalysed reaction, which then generates a signal. However, analysts are also concerned with the detection of large structures such as foreign proteins or toxins, invading bacteria or viruses, and these are clearly outside the normal scale of participants in analytical enzyme reactions. Fortunately, an alternative set of detectors with comparable high specificity is available in the form of the immunoglobulins (antibodies).

The immune system is a powerful body defence mechanism designed to detect and destroy antigen-bearing entities such as bacteria, invading viruses or foreign macromolecules. Antibodies play a key role in the detection of antigens, and in the promotion of antigen destruction. The antibodies are produced by special cells in response to the presence of a given antigen, and antibodies so produced bind specifically to that particular antigen target. This remarkably high degree of specificity in the binding of an antibody to its antigen provides the basis for an analytical procedure called the immunoassay.

The reaction of antibody binding to antigen provides the necessary specificity for accurate detection of the antigens, but the reaction requires a sensitive and specific method of signal generation for quantitative analysis. A suitable transducer in this case can be provided by the addition of an enzyme to produce the enzyme immunoassay (EIA) (see Wilson and Goulding, 1986). The enzyme can be immobilized onto the antigen to form an enzyme-antigen complex in which the enzyme retains its catalytic activity. A simple account of the basic EIA procedure is outlined in Fig. 3.9.

The method depends on the antibody-antigen interaction causing inhibition of the enzyme catalysed reaction. Thus the first step in an EIA is a control experiment to establish the level of enzyme activity associated with the enzyme-antigen complex. This is accomplished by supplying substrate for the enzyme, and monitoring the formation of product. In the test reaction, the enzyme-antigen complex is mixed with the unknown sample containing some free antigen, then a *limited* amount of antibody is added to the mixture. An amount of substrate is added for the enzyme, and residual enzyme activity is estimated by monitoring the formation of product. The antibody will bind to both the free antigen, derived from the unknown sample, and to the antigen in the enzyme-antigen complex.

The degree of enzyme inhibition will depend on the concentration of free antigen in the sample. If the concentration of free antigen is low or nil, then almost all of the antibody will bind to the enzyme-antigen complex, causing a substantial reduction in the level of residual enzyme activity. If the concentration of free antigen is high, then most of the antibody will bind to the free antigen, and, correspondingly, the level of residual enzyme activity will be high. By employing careful controls and standards it is possible to produce an accurate calibration to relate the level of residual enzyme activity to the concentration of free antigen present in the unknown sample.

To be suitable as a signal generator in an EIA procedure, an enzyme and the reaction that it catalyses must meet several important criteria. For example, it is essential that the enzyme assay procedure should be inexpensive, simple to perform, easy to monitor and sensitive for detecting low levels of enzyme activity. In addition the enzyme must not be a normal constituent of biological fluids such as blood, and the components of the enzyme reaction (substrates and products) should not be normal constituents of biological fluids. Further, the enzyme should be available cheaply in a pure form, have a high catalytic activity, a high degree of substrate specificity, and good stability characteristics. Since no one enzyme is available that would meet all of the criteria, there are several different enzymes that have been used frequently in enzyme

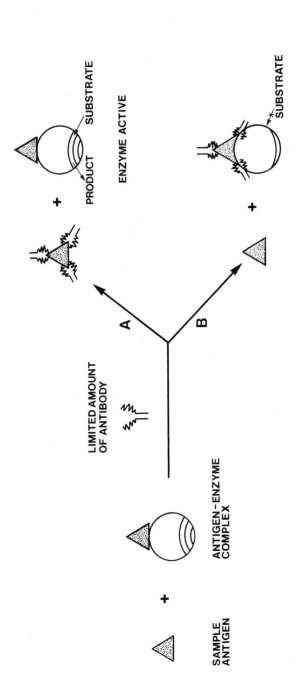

Fig. 3.9 Schematic illustration of a simple enzyme immunoassay procedure. Path **A** is predominant if the concentration of antigen in the sample is high. Path **B** is predominant if the concentration of antigen in the sample is low.

SAMPLE ANTIGEN

ANTIGEN–ENZYME COMPLEX

LIMITED AMOUNT OF ANTIBODY

A

B

SUBSTRATE

PRODUCT

ENZYME ACTIVE

SUBSTRATE

ENZYME INACTIVE

immunoassays. A few of the enzymes that have found use in EIA techniques are peroxidase, glucose oxidase, alkaline phosphatase, beta-galactosidase and lysozyme.

The EIA has been used to measure a wide range of molecules including hormones such as thyroxine, insulin, progesterone and various oestrogens, bacterial toxins such as cholera exotoxin, and serum proteins such as albumin. An enzyme immunoassay is used extensively as a diagnostic test for pregnancy in women. The test is based on the detection of a hormone called human chorionic gonadotropin (HCG), which is produced shortly after conception. Using anti-HCG antibodies and an HCG-enzyme complex to detect the presence of HCG in urine, it is possible to confirm a pregnancy three to four weeks after conception as the level of HCG at this stage is high enough for reliable detection.

A very useful extension of the basic EIA technique is provided by simple rearrangement of the procedure to enable the detection of specific antibodies in blood serum. By detecting the presence of a specific antibody this, by implication, identifies the specific organism or molecule causing the production of the antibodies. Thus this procedure can provide a powerful tool in the diagnosis of disease. For example, it is sometimes difficult to distinguish between the various forms of measles disease. Using an EIA to detect the antibodies against Rubella virus can facilitate the diagnosis of German measles. To further demonstrate the potential of the enzyme immunoassay in the diagnosis of disease, it will be useful to illustrate the significance of a recent enzyme immunoassay for detection of leprosy (see Mackensie, 1984).

Leprosy is caused by a micro-organism called *Mycobacterium leprae*. The micro-organism is transmitted between humans and resides in the skin and nerves, where it often produces a loss of feeling or sensation as nerve function is disrupted. *M. leprae* is a successful parasite, and does not aim to kill its host. However, the leprosy patient may develop skin deformities such as nodules or lumps, due to the growth of the organism under the skin. The loss of feeling in the skin areas of the hands and feet is a serious complication, which often leads to skin abrasion or breakage, and secondary bacterial infections. Without feeling to guard against mechanical damage, the skin is particularly vulnerable. A nurse once recalled that on leaving the house of an infected family, she found difficulty turning the key in a rusty old lock. A young leprosy patient came to her assistance and turned the key to open the door. The nurse examined the child's hand, and found that the skin was broken and bleeding. The child was not able to feel the pain as the key broke through the skin on his hand. Also, in the poorer, rat-infested leprosy colonies, it is not unknown for patients (while asleep) to lose fingers or toes – because of the absence of feeling, the patient is unaware of the rat gnawing through the skin.

Fortunately, leprosy is controllable and an effective drug called Dapsone (Fig. 3.10) has been available for over forty years. Yet the incidence of the

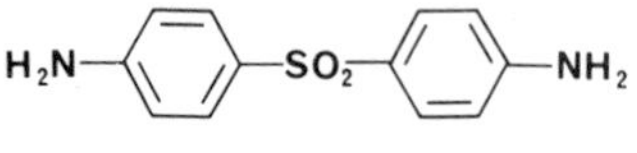

$$H_2N-\langle\bigcirc\rangle-SO_2-\langle\bigcirc\rangle-NH_2$$

Fig. 3.10   Structure of Dapsone, anti-leprosy drug.

disease has not declined, and the number of people suffering from the disease has stayed at around ten million, mainly in Southern Asia and Africa. It is now recognized that patients can carry the disease, and spread it for years without knowing it. If leprosy is to be eradicated then it will be essential to cure existing cases, *and* prevent the spread of the disease. Since a carrier of the disease shows none of the obvious symptoms, it has so far proved impossible to detect carriers and thereby check the spread of the disease.

After many years of intensive research, an enzyme immunoassay has been developed which will be of considerable importance in the detection of people who are, unknowingly, carriers of leprosy. The method is a slightly modified enzyme immunoassay called an enzyme linked immunosorbent assay (ELISA). The procedure is designed to detect the occurence in the blood of the specific antibodies produced in response to the presence of *M. leprae*, which thereby implies the existence of the disease in the body. A leprosy antigen-enzyme complex is trapped on a thin layer of gel material, which is coated on the surface of a small dish. When blood serum containing anti-leprosy antibodies is applied to the dish, the antibody binds to the antigen-enzyme complex causing inhibition of the enzyme. The enzyme reaction is designed to produce a coloured product. A positive test is indicated by little or no colour production due to the inhibition of the enzyme-antigen complex by the antibodies.

It is hoped that this ELISA will enable the medical authorities to screen large numbers of people in areas where leprosy is prevalent, and thereby detect both leprosy carriers and patients in the early stages of the disease. The test may also prove very useful in a variety of clinical investigations to establish how the disease is transmitted, established in the skin and responds to the drug therapy. In particular it is apparent that the organism *M. leprae* can successfully deceive the body defence mechanisms in order to maintain its residence in the skin and nerves. The test may provide a useful tool for elucidating the molecular basis responsible for this clever deception.

The general use of antibodies in analysis has been restricted because of the high cost of producing fairly small amounts of pure antibody. Until recently, antibodies were isolated from the blood of previously immunized animals, by intensive methods of extraction and purification. However, the recent advances in monoclonal antibody research will enable large scale production of a wide range of useful antibodies. This development will lead to a significant increase in the use of enzyme immunoassay procedures for analytical work.

3.7 Food quality control

Analysis of foodstuffs has developed considerably since it was first proposed in 1860 as a means of detecting the adulteration of food with toxic additives. It was common, in those days, to use copper salts, red lead and various arsenic pigments as colouring agents in food preparations, with frequent fatal consequences. Also it was not unusual for unscrupulous people to produce and sell table vinegar made from sulphuric acid. Fortunately, legislation now restricts the adulteration of foodstuffs, and modern analysis performs a vital role in the maintenance and improvement of high standards in the preparation of foodstuffs.

A large section of the food industry is based on the preparation and processing of largely perishable material derived from plants and animals. In many cases the foodstuff consists of dead tissues, organs or organisms at various stages of disintegration. It is therefore important to monitor the integrity of foodstuffs as they pass through the processing stages to ensure that the end-product is both safe and appealing to the customer. Enzymes have been applied to the analysis of a wide range of substances in foodstuffs, from the level of alcohol in wine to the concentration of residual insecticide in plant tissues. Also the assay of endogenous enzymes can serve as a useful indicator of the success of a particular food processing step. The efficiency of the pasteurization of milk can be estimated by measuring the activity of the enzyme invertase before and after pasteurization. Effective pasteurization should produce substantial inactivation of the enzyme. In relation to the maintenance of high standards in the food industry, enzymes are used to detect the adulteration of food products with inferior quality substitutes. A representative indication of the application of enzymes in food analysis is provided by considering two methods used in meat analysis. Firstly, the use of an enzyme electrode in the estimation of meat freshness, and secondly the use of an ELISA method to detect the adulteration of beef meat with meat substitutes from other sources.

One of the fundamental characteristics associated with life is the intricate control of the complicated mechanisms of metabolism in cells and between tissues. In life there is a dynamic balance between the degradation and the biosynthesis of macromolecules to ensure a continual supply of basic components for new and replacement growth. After death, the control procedures cease, and since biosynthesis requires coordination of component supply and an input of energy, it also ceases to function. Also, the enzymes responsible for degradation, which are normally under rigorous metabolic control, are released from metabolic restraint. These enzymes then begin a steady breakdown of tissues and cells to produce a variety of degradation products such as ammonia, amines, carboxylates and carbon dioxide, as biological macromolecules are broken down to simpler components. The accumulation of simple molecules provides an ideal environment for the growth of bacteria, yeast and fungi. These organisms promote further deterioration and spoilage of meat, producing an alteration in the colour, flavour, texture and eventually the odour of the meat. Since the ingestion of bacteria may cause food poisoning, meat that has been allowed to deteriorate beyond strict limits is deemed not fit for human consumption. Therefore, estimation of meat freshness is very important in the food industry for the safe processing and preparation of high quality meat products.

The most common procedure for the estimation of meat freshness is a chemical determination of NH_3 following acid hydrolysis of total organic nitrogen (Kjeldahl analysis). This lengthy procedure involves several steps, including extractions, distillations and titrations. There is, therefore, a good opportunity for the introduction of a simple and reproducible method involving an enzyme. Since various kinds of amines are produced in the disintegration of meat, monitoring of these substances may provide a useful indicator of meat freshness. Monoamines such as histamine and tyramine arise in meat from the decarboxylation of amino acids, which themselves have been released by the proteolysis of meat protein.

The monoamines can be measured using an enzyme electrode system consisting of the enzyme monoamine oxidase and an oxygen electrode as transducer.

$$\underset{\text{(monoamine)}}{RCH_2NH_2} + O_2 + H_2O \xrightarrow{\text{monoamine oxidase}} \underset{\text{(aldehyde)}}{RCHO} + H_2O_2 + NH_3$$

The oxygen electrode transducer monitors the decrease in dissolved oxygen concentration as the monoamine is oxidized. The electric current generated is directly proportional to the initial concentration of monoamines. The enzyme is immobilized on a simple membrane, which is attached to the oxygen electrode. In comparison with the conventional Kjeldahl nitrogen analysis, the enzyme electrode provides a simple, reliable and economic method for the routine analysis of monoamines to estimate meat freshness.

The adulteration of beef meat with meat from other animals or soya protein is difficult to detect, especially if the meat is minced or processed. However, the recent application of an enzyme-linked immunosorbent assay (ELISA) procedure will allow meat analysts to distinguish between cattle, horse, pig and sheep meat, and soya protein. The analysis is based on the fact that animals produce antibodies in response to the presence of foreign protein. Cattle are used to 'raise' the antibodies against meat proteins from the main sources of substitute material used in adulteration of beef, i.e. horse, sheep, pig meat and soya protein. The ELISA procedure is outlined in Fig. 3.11. In this test, the suspect meat is extracted into solution. Then the meat protein solution is spread over a layer of gel on a small plastic dish. The gel is designed to adsorb the protein from the solution. After washing, a layer of protein derived from the meat is retained on top of the gel. The next stage is to determine if any of the protein is foreign, i.e. not from cattle, using antibody-enzyme complexes. If it is decided to test for horse meat protein, then the anti-horse meat protein antibody-enzyme complex is layered on top of the protein layer. The antibody will recognize and bind to horse meat protein on the gel. After washing, to remove excess antibody-enzyme complex, enzyme substrate is added. If the antibody-enzyme complex has been retained, i.e. horse meat protein present, then the enzyme will catalyse the formation of a coloured product using the added substrate.

The technique is very sensitive, and can detect adulterating substitutes at levels as low as 5%. The test is designed for analysis of raw meat, but has now been extended for analysis of cooked meat. This application will be of great use to the meat analysts faced with a large and expanding array of prepared meat products such as steaklets, mince pies, beef curry, beef sausages and hamburgers.

In addition to testing meat products, the EIAs have many other applications in the analysis of food products. It is possible to differentiate milk and cheese products from cow, goat and sheep, and thereby detect if a goat milk product

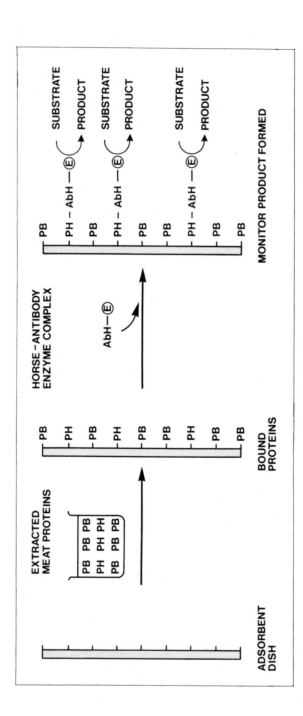

Fig. 3.11 Use of an ELISA procedure to detect the adulteration of beef meat protein (PB) with horse meat protein (PH).

has been invested with some cow milk. The quality of chocolate sweets can be checked to detect the substitution of cocoa by the cheaper carob powder. Also, a number of food intolerances and food allergies necessitate the screening of food products for the causative agent of the intolerance or allergy. For example, in coeliac disease the agent is gliadin, which is one of the proteins present in the endosperm of wheat. In patients with this disease the gliadin causes atrophy of the intestinal mucosa villi, and the consequences vary from life threatening to mild cases involving an upleasant form of diarrhoea. An EIA for gliadin makes it possible to screen food products, and eliminate from the diet those products containing high levels of gliadin.

3.8 Environmental analysis

During the rapid technological advances of the last 300 years, man has steadily acquired the desire and the power to change his environment radically by replacing green land with concrete cities, and by removing natural communities and replacing them with crops and farm animals. Unfortunately, although much innovation and invention went into making the changes in the environment, man has shown less insight and concern for protecting the environment from the harmful side effects, such as pollution.

Most pollutants are biodegradable and are harmful for only a short period until they are broken down by the scavenging components of the environment. For example, sewage discharged into the sea is eventually rendered harmless by bacteria and other organisms. However, some pollutants retain their harmful characteristics for a long time, and these present a more serious environmental problem because of the danger of them accumulating through the natural food chains. Lead and mercury are well known persistent, accumulative pollutants, as are the polychlorinated biphenyls (PCBs). Environmental analysis is concerned with the detection and evaluation of temporary and persistent pollutants so that the harmful effects may be quantified and corrective controls brought into operation to reduce the level of the pollutant in the environment.

Enzymes are particularly suitable for environmental analysis because the toxicity of most pollutants is related to their ability to inactivate important enzyme systems in the body. Many of the organic insecticides are designed to kill insects by inhibiting cell respiration and ATP synthesis. Others, particularly the chlorinated organic compounds such as DDT act on the insect nervous system. Thus the detection of enzyme inhibition can provide a useful method of monitoring pollutants in the environment. This method depends on the reaction of the enzyme (E) with the inhibitor (I) to produce a catalytically inactive complex (EI).

$$E + S \rightleftharpoons ES \rightleftharpoons E + P \qquad E + I \rightleftharpoons EI \xrightarrow{\quad\times\quad}$$

As the number of active enzyme molecules decreases the conversion of substrate (S) to product (P) shows a corresponding decrease.

A particularly useful candidate for this type of analysis is the enzyme

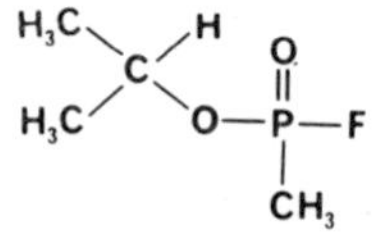

SARIN, NERVE GAS

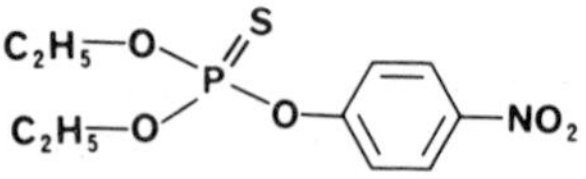

PARATHION, INSECTICIDE

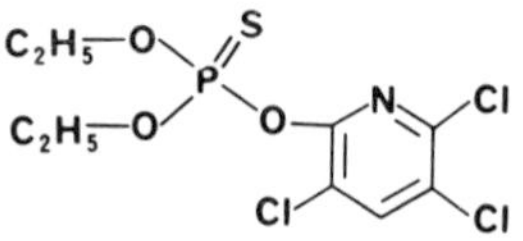

DURSBAN, INSECTICIDE

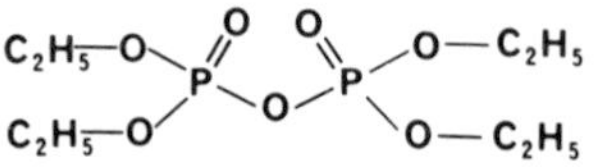

TETRAETHYLPYROPHOSPHATE
(TEPP), INSECTICIDE

Fig. 3.12   Some inhibitors of cholinesterase.

cholinesterase. This enzyme is inhibited by a large number of organic compounds including many insecticides and some of the nerve gases developed for chemical warfare (Fig. 3.12). Several biosensor instruments are commercially available for monitoring air and/or water samples. The enzyme is normally immobilized in a gel matrix, and layered onto a small polyurethane pad – suitable substrate for the enzyme is butyrylthiocholine iodide. During a test, the enzyme impregnated pad is manoeuvred into the path of a flow of substrate solution, which the enzyme readily hydrolyses to form the product, thiocholine iodide.

$$\text{butyrylthiocholine iodide} \xrightarrow{\text{cholinesterase}} \text{thiocholine iodide + butyric acid}$$

The enzyme pad is situated between the anode and cathode of a potentiometric electrode, which functions as a transducer in this system. The thiocholine iodide is oxidized at the anode of the electrode giving rise to a potential of about 250 mV. Thus in the absence of inhibitor, a steady voltage of around 250 mV is recorded. The sample to be tested is then mixed in with the substrate flow. If the sample contains a cholinesterase inhibitor, the enzyme will lose activity. The level of thiocholine iodide will decrease, and the electrode potential increases to around 500 mV. This change in potential can be related to the concentration of cholinesterase inhibitor present in the sample. The method is sensitive and can detect microgram quantities of insecticides. Also, the method is rapid so a number of samples may be analysed in a short space of time. However, the system does not detect pollutants that do not cause inhibition of cholinesterase, and the system cannot identify unknown pollutants. The system can be readily adapted for monitoring both air and water samples. In each case this can be done continuously, using an automatic monitoring instrument, which may be connected to a suitable alarm mechanism.

Various other enzyme-based biosensors are being evaluated for possible application in the estimation of sewage contamination on sandy beaches, and in assaying the activity of soil microflora for determining the quality of different soils. Several enzyme systems have been applied to the detection of pollutants such as phenol, phenoxyacids such as the hormone-analogue herbicide 2,4-D (2,4-dichlorophenoxyacetic acid), nitrilotriacetic acid (component of some detergents), and heavy metals such as copper, zinc and lead (see Russell, 1984).

3.9 The microbiosensor chip

Research into biosensor design is making steady progress, and although there are many serious problems that have still to be overcome, the future could present some exciting developments in the fundamental design of biosensors. One of the main problems associated with the application of biosensors is the fact that almost all of the current biosensors are mono-functional, i.e. each biosensor measures only one analyte. Given the large number of analytes that are measured, it is clear that it would not be practical or desirable to maintain a correspondingly large number of biosensors. It is now recognized that there is a very strong case for the design and development of multi-functional biosensors that are capable of measuring a minimum of ten different analytes.

The successful development of multi-functional biosensors will require miniaturization of key biosensor components, and the use of sophisticated data handling equipment for the greater amount of information that will be generated. The most likely solution for the latter problem is the microprocessor, which is a relatively inexpensive computer. The microprocessor is ideally suited for assimilation, manipulation and storage of the large quantities of biochemical data that would be generated by a multi-functional biosensor. Indeed, the use of microprocessors independently or incorporated into scientific equipment has already facilitated a substantial increase in data processing associated with scientific instrumentation.

The successful miniaturization of the biosensor is an important factor in the design of a multi-functional instrument. The basic components are the enzyme and the transducer. Since enzymes are already minute in size (average 10 nm in diameter), it is the transducers that require miniaturization. A possible line of approach may again be provided by the electronics industry, which has developed the ability to produce very small, highly complex structures, with good reliability and at low cost. In particular, field effect transistors (FET), which are semi-conductor devices used in electronic circuitry, have been adapted to function as semi-conductor transducers, for example ISFET (ion selective) and CHEMFET (chemically sensitive). For more details on CHEMFET see Sibbald (1983). These are close relatives of the silicon chip in which a charge flowing through an electronic gate is altered by applying a second charge to it. In a CHEMFET, the main charge is altered by a change in the concentration of ions. A membrane that binds certain ions is attached to a specific part of the CHEMFET. When the CHEMFET is introduced into a

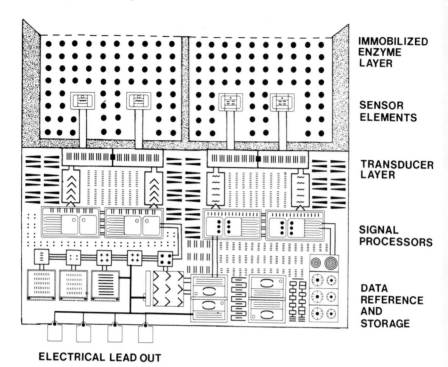

IMMOBILIZED
ENZYME
LAYER

SENSOR
ELEMENTS

TRANSDUCER
LAYER

SIGNAL
PROCESSORS

DATA
REFERENCE
AND
STORAGE

ELECTRICAL LEAD OUT

Fig. 3.13   A representation of some of the desirable features that may be incorporated in a multi-functional microbiosensor chip.

solution containing these ions, the ions bind to the membrane. The local concentration of ions in a particular position is sufficient to cause a fluctuation in the current flowing through the device. Like the silicon chip, the CHEMFET is very small, approximately 2 mm square.

Perhaps the most exciting prospect from the research work on biosensors is the incorporation of enzymes into FETs to produce miniature enzyme electrodes. It seems feasible that current microelectronic microfabrication technology may be adapted to construct electronic chips modified to function as specific biosensors. A hypothetical microbiosensor chip is depicted in Fig. 3.13. It is envisaged that such a chip would be composed of semi-conductor material, structured to provide small channels and gates for sample flow. The device would also contain small compartments for locating immobilized enzymes, miniature sensors and associated electronics. The eventual aim would be the construction of an instrument containing an array of biosensor chips, incorporating a variety of immobilized enzymes, capable of measuring a number of analytes simultaneously, and employing the assistance of a microprocessor to manage the data handling. Such a development could lead to a new generation of analytical multi-biosensors for use in medicine, industry, agriculture and environmental analysis. In medicine, this could enable a wider range of clinical tests to be performed by general practitioners and nurses. In addition the miniaturization may enable the *in vivo* measurement of key metabolites on occasions when the monitoring of vital body signs is critical, for example during major surgery, in intensive care, and during temporary paramedical treatment at the site of an accident.

With a multi-million pound market in view, there is worldwide interest in the development of these devices. A few studies have been widely reported, for example demonstrating an FET sensitive to penicillin, but the intense commercial competition surrounding research work in this field is likely to ensure that new devices are kept 'top secret' until patents have been secured.

4

Enzymes in Medical Therapy

4.1 Introduction

The use of enzymes in clinical therapy has been known for over 100 years since early surgeons used crude preparations of protease enzymes as surgical aids. In more recent times, enzyme therapy has been proposed for the alleviation of a number of medical disorders. The underlying basis for this proposal has been the substantial advances in biochemistry, which have provided a greater understanding of cell chemistry and metabolism. This in turn has illuminated the role of enzymes in the molecular and metabolic dysfunctions associated with many diseases.

The present level of enzyme applications in medical therapy has so far not matched the considerable potential that is widely recognized as being reasonably possible. The relatively slow development of enzyme therapy can be attributed to several limiting factors, which are mostly technical problems rather than inadequacies in the basic concept. Some of these problems are identified below.

(1) Repeated injections of a foreign enzyme into the body will trigger the immune system into action.

(2) Many enzymes are, as yet, not available in a pure form, and in sufficient quantity at low cost.

(3) In some cases the required enzyme is needed in fairly inaccessible parts of the body, hence it may be difficult to deliver the enzyme to the target site.

(4) Some enzymes have inherently poor stability.

(5) Enzymes injected into the blood are readily inactivated by body defence systems.

(6) Some enzymes have very specific cofactor requirements, and if the cofactor is not available naturally at the target site then it must be supplied together with the enzyme.

In order to fulfill the great expectations of enzyme therapy, the above fundamental problems and potential hazards must be thoroughly investigated and eliminated.

The immune system is triggered by the presence in the body of a biochemical entity that the body recognizes as foreign. That entity may be living (e.g. bacterial cells) or non-living (e.g. a foreign enzyme). When the system is activated the body organizes a wide array of cells, proteins and enzymes to attack the foreign body. The gross symptoms associated with a strong immune reaction are fever, swollen glands, skin inflammation or rash, vomiting and diarrhoea. The body defences are characteristically single-minded, and strong reactions provoke a fight to the death, either of the body or of the invader, so a severe immune reaction in a weak patient can be fatal. There are a number of immunosuppressive drugs that can be used in conjunction with enzyme therapy. These drugs check the production of antibodies and leucocytes, rendering the patient immunologically defenceless and vulnerable to almost any disease.

The enzyme production industry is well developed for the isolation and purification of enzymes from bacterial sources, and therefore bacterial enzymes are more readily available. Although many bacterial enzymes can be produced in a pure form, free from contamination with toxins, the bacterial enzyme is nevertheless recognized as foreign and invokes a serious immune response when repeatedly injected into the body. Enzyme isolation and purification from human or animal sources is limited, and consequently fewer pure enzymes are available at low cost. A possible source of human enzymes is human placenta, which is available in fairly large quantities, and research efforts are evaluating the potential for large scale enzyme isolation from placental tissue. Looking to the future it is likely that genetic engineering techniques will lead to the production of human enzymes by fermentation using micro-organisms.

Effective enzyme therapy requires contact between the enzyme and its intended substrate, and success in arranging efficient contact depends on the location of the substrate in the body. As shown in Fig. 4.1, if the substrate accumulates in an extracellular location, for example blood vessel, then it is comparatively simple to deliver enzyme to the site of accumulation. However, if the substrate accumulates in an intracellular position such that the enzyme must be delivered inside cells, then the difficulty involves first transporting the enzyme to the target cell, and secondly promoting specific uptake of the enzyme into these cells. In some cases the substrate may be located in a subcellular organelle, for example lysosome or mitochondrion, and this extra level of difficulty further complicates the problems associated with delivering the enzyme to the target site.

Enzymes generally have rather poor stability characteristics and can readily lose catalytic activity if the sensitive protein structure is disrupted. Also, loss of activity can occur if the enzyme is attacked by body defence proteins and cells. The stabilization and protection of the enzyme can usually be accomplished in one step by immobilizing the enzyme within a semi-permeable membrane as illustrated in Fig. 4.2. It is essential to select a semi-permeable membrane material that will not invoke an immune response, and such a membrane may be constructed using proteins or lipids obtained from the blood or body fat of the patient. Another approach employs the 'Trojan horse' trick, using red blood cells that can be emptied and then refilled with enzyme by osmotic shock techniques. When these cells are removed from the patient, filled with enzyme and returned to the patient, no immune response is detected. Also, worn out red

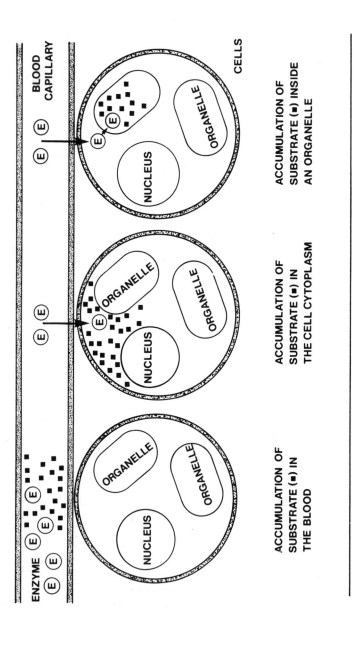

Fig. 4.1 Diagram showing the possible levels of difficulty associated with specific delivery of a replacement enzyme to a target site (see text for further discussion).

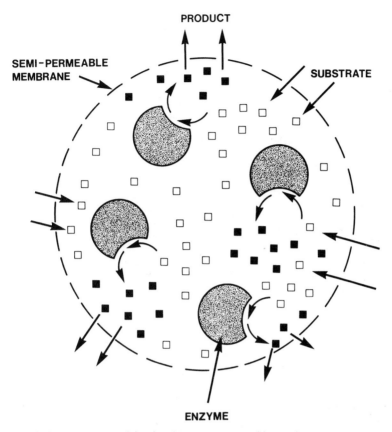

Fig. 4.2   Enzyme immobilized within a semi-permeable membrane.

blood cells are removed from circulation by the liver, so the red blood cell provides specific vehicle for delivering therapeutic enzymes to the liver.

   Although enzyme therapy is still in its infancy, and there is a great deal of work still to be done, the outlook for the future is very promising indeed. It is hoped that the following discussions will give the reader a clear insight into the principal areas of enzyme application in clinical therapy.

4.2   Enzyme replacement

Over 125 human diseases are known that can be attributed to inherited errors in biochemical metabolism, arising from some form of enzyme deficiency disease. In such medical disorders, the dysfunction is caused by the absence or inactivation of an enzyme normally present in the body. A short list of some of the medical disorders due to enzyme deficiency is given in Table 4.1. The most logical therapeutic approach is to replace the missing or inactive enzyme and restore the normal metabolic conditions. The application of enzyme

Table 4.1   Some diseases caused by enzyme deficiency.

Disease	Enzyme deficiency
Acatalasaemia	Catalase
Alcaptonuria	Homogentisate 1,2-dioxygenase
Christmas disease	Factor IX
Citrullinaemia	Arginosuccinate synthetase
Fabry's disease	Alpha-galactosidase
Fucosidosis	Alpha-fucosidase
Galactosaemia	UDPG-hexose-1-phosphate transferase
Gaucher's disease	Beta-glucocerebrosidase
Haemophilia A	Factor VIII
Lesch-Nyhan syndrome	Hypoxanthine-guanine phosphoribose transferase
Nieman-Pick disease	Sphingomyelin phosphodiesterase
Pentosuria	Xylulose reductase
Phenylketonuria	Phenylalanine hydroxylase
Von-Gierke's disease	Glucose-6-phosphatase

replacement in medical therapy has so far been limited to only a few notable cases, and although enzyme therapy has the potential for much greater application, the slow progress reflects the considerable difficulties that are still under investigation.

As noted earlier, a key problem that can be simple or very difficult to solve is delivery of the appropriate enzyme to the specific target site. This depends on the accessibility of the cells in need of the replacement enzyme as illustrated in Fig. 4.1. In cases involving a deficiency of digestive enzymes, it is relatively simple to deliver replacement enzyme to the target site (the gut). For example, such enzymes are used to help alleviate some of the symptoms associated with cystic fibrosis. This is an inherited disease affecting about 1 in every 1000 newborn babies, and one of the main complications is a degeneration of the pancreas, causing loss of the pancreatic enzymes. These enzymes are essential for the digestion of proteins and fats (in foodstuff) to simpler components that are suitable for absorption by the digestive system. The basic therapy consists of supplying patients with high potency extracts of bovine pancreas in capsules or tablets. The treatment is not wholly satisfactory because a substantial percentage of the replacement enzymes are destroyed by the acid in the stomach. New developments using pancreatic enzymes immobilized for greater stability are under investigation, with the aim of improving the efficiency of enzyme use and the residence time in the gut.

The specific delivery of replacement enzyme becomes more complicated when access to the target cells is restricted. The extent of the problem varies but in some cases, for example the Lesch-Nyhan syndrome, the delivery of enzyme represents a formidable obstacle. This is a very intriguing enzyme deficiency disease, because the biochemical defect gives rise to an alteration in the emotional behaviour of the patient. The syndrome is a severe neurological disease occurring in males, which is characterized by involuntary writhing motions, mental retardation and a compulsion for self-mutilation of lips and fingers by biting. To avoid permanent disfigurement, many patients must wear arm restraints to prevent them from biting off their fingertips. In severe cases, removal of all teeth is required to prevent loss of parts of the tongue and lips. The disease is caused by the absence of an enzyme called hypoxanthine-guanine phosphoribosyl transferase (HGPRT). This enzyme is involved in the salvage of purine bases released by nucleic acid breakdown, and clearly plays a key part in nucleotide metabolism. Most body cells have a back-up system to accomplish the same function as that performed by HGPRT, and therefore are not greatly disturbed by the absence of this enzyme. Unfortunately, the cells of the basal ganglia of the brain do not possess such a back-up system, and without HGPRT the metabolism of nucleotides and nucleic acids in these brain cells is seriously impaired. With regard to enzyme therapy, a suitable source of relacement enzyme is available. However, the difficulties involved in delivering the replacement HGPRT across the very protective blood-brain barrier to reach the target brain cells are considerable, and present an immense challenge to the research scientists working in this area. For a recent review on the blood-brain barrier see Goldstein and Betz (1986).

Various vehicles are being evaluated for transporting replacement enzymes through the bloodstream to specific target cells. One interesting system that is receiving considerable attention is the liposome. This is a capsule composed of lipid bilayers alternating with aqueous layers into which the replacement enzyme is trapped by encapsulation. An illustration of the liposome is given in Fig. 4.3. When liposomes are injected into the bloodstream, they are selectively removed by the liver and the spleen, and therefore present a specific route to these tissues. Further, research work indicates that it may be possible to modify the surface properties of liposomes to promote specific uptake by other tissues. It is known that cells recognize each other by the cell surface markers that they possess, and it is hoped that liposomes may be programmed for a specific tissue by incorporating a specific set of markers on the surface of the liposome. For example, the incorporation of phosphomannose (marker) onto the surface of liposomes promotes the uptake of these liposomes by fibroblast cells of connective tissue. Although more research is required to decipher the cell-surface marker codes, it is clear that this approach could have a substantial impact on the problem of enzyme delivery to the target site.

One particular drawback that slows the advance of research in enzyme replacement therapy is the lack of suitable animal models for studying the effectiveness of therapeutic treatments for application in human diseases. When no comparable model is available, then it is necessary to construct an artificial system for testing a new therapeutic development. Although such systems can often give a good indication of the likely level of success, the full implications of any new development will only become apparent when applied to the human patient. In a few cases where a comparable animal model exists, the development of enzyme therapy has proceeded faster, because the operating problems are encountered and solved in advance of any clinical tests on human patients.

One animal model that has been thoroughly studied is Feinstein's strain of mice, which have a hereditary deficiency of the enzyme catalase. In this strain of mice, blood catalase activity is only about 2% of normal, and total body

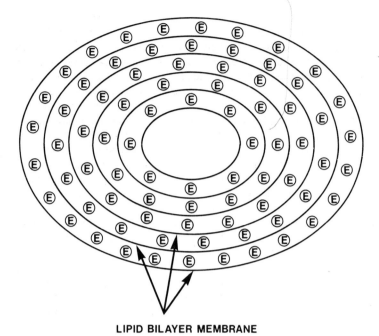

LIPID BILAYER MEMBRANE

Fig. 4.3 Immobilization of an enzyme by entrapment in a liposome. The lipid layers are separated by aqueous layers containing soluble enzyme.

catalase is about 20% of normal, which is just about sufficient to enable survival. Catalase is an important enzyme in the body responsible for eliminating potentially harmful hydrogen peroxide (H_2O_2) from the blood and tissues.

$$H_2O_2 \xrightarrow{\text{catalase}} 2H_2O + O_2$$

A replacement therapy has been developed using catalase from bovine liver, immobilized by entrapment in microcapsules. It involves injection of the small microcapsules into the peritoneal cavity, and follow-up studies have shown that the enzyme functions efficiently in reducing and maintaining H_2O_2 at a safe low level. The treatment allowed the mice to return to a near normal existence, and no serious adverse reactions were detected as a result of the therapy. The application of a similar system to human patients with this disease is under investigation, and results appear encouraging (see Chang, 1977).

An important prerequisite for the successful application of any enzyme replacement is a thorough understanding of the biochemical implications of any added material (containing enzyme), and its consequent effect on the local and wider metabolic environment. It is essential that the fundamental biochemistry is apparent so that the full implications of a therapeutic strategy may be known, and new applications devised that are safe and compatible with the normal metabolism of the patient. However, the intense pressure for therapy sometimes leads to the adoption of an enzyme replacement strategy before all the basic investigations are complete. Then as the fundamental research catches up, it may become evident that the original strategy is not completely safe. This problem has become apparent in the use of blood plasma for the treatment of deficiences in blood clotting. It is now clear that blood plasma is not a completely safe source of replacement enzymes, and on occasion it may contain harmful constituents.

The formation of blood clots is a complicated process with a high level of control built into the mechanism. Such a sophisticated system is essential in view of the fine balance between haemorrhage (bleeding) and thrombosis (excessive clotting), given the need for clotting to occur rapidly at the site of injury. The process involves many proteins called Factors, which are given Roman numerals for ease of identification. Most of the Factors are protease enzymes that are present in the blood as inactive precursors. Clotting is initiated by activation of the first of these Factors. This Factor then starts a cascade of reactions in which one enzyme catalyses activation of the next in the sequence, which then catalyses activation of the next and so on. In the final stage, inactive prothrombin is converted to active thrombin by Factor X, and thrombin catalyses the activation of fibrinogen to produce the protein fibrin for clot formation (see the right-hand side of Fig. 4.4).

The most common congenital defects are haemophilia A (lack of Factor VIII) and haemophilia B (lack of Factor IX), and it is estimated that in Britain there are about 4000 patients with haemophilia A disease. Generally, therapeutic treatment involves enzyme replacement by supplying a blood plasma fraction containing the protease Factors. The fraction is extracted from whole blood donated by the general public. This treatment substantially improves both the survival and the quality of life for haemophiliacs. For example, children are able to take part in sports activities without the danger of excessive bleeding in the joints, which normally disables them before they reach early adulthood. However, as research work into the composition of the plasma extract has progressed, it has become clear that blood donated by the general public may contain contaminants that persist in the plasma extracts used for haemophilia therapy. In particular, there is a risk that the donated blood may contain hepatitis B virus, which if passed on to the haemophiliac could cause chronic liver disease.

More recent research has found that some plasma extracts may contain a virus that attacks important cells which play a vital role in the immune system. The virus is called human thymus-cell lymphotrophic virus or HTLV-III for short, and it selectively invades this particular type of white blood cell. After invasion, normal cell activities cease, viral particles replicate, and the cells are eventually destroyed to allow release into the blood of large numbers of new virus particles. Since the virus effectively destroys a vital link in the immune system, the body defences become deficient, and any subsequent infection may prove fatal. Plasma extracts contaminated with HTLV-III virus may communicate the autoimmune disease known as AIDS (acquired immune

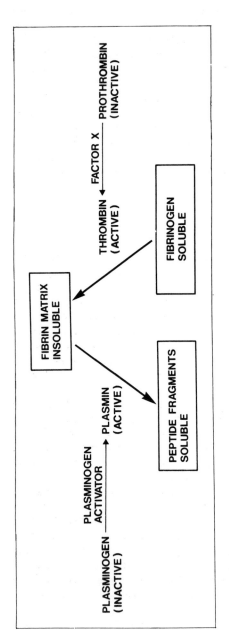

Fig. 4.4 Enzyme catalysed formation (thrombin) and dissolution (plasmin) of fibrin clots.

deficiency syndrome) to haemophiliac and other patients receiving blood-related products.

Another problem with the use of plasma extracts is that the shelf-life of the extracts is short, because some of the enzyme Factors are not very stable in this preparation. Several lines of research are aiming to overcome the problems associated with the use of blood plasma extracts. In the short term a method has been tested for sterilizing plasma extracts by heating at 68°C for 24 hours. This high temperature has been found to destroy the viruses such as HTLV-III without affecting the important Factor VIII. A test has been developed for indicating the presence of AIDS virus in blood samples, and is now in routine use for screening blood donations and assisting with diagnosis of AIDS disease. The test involves an enzyme immunoassay procedure (see section 3.6). Other studies are concerned with the extraction, purification and immobilization of the individual enzymes, such as Factor VIII. In the future, it is likely that genetic engineering techniques will enable the manufacture of large quantities of pure Factor VIII, and provide a safer enzyme replacement therapy for haemophilia patients.

4.3 Cancer treatment

Cancer is a disease that can affect all multicellular organisms, and can strike almost anywhere in the human body; although certain sites are especially susceptible, for example mouth, skin, respiratory organs, blood, lymph, digestive organs, urinary organs, breasts and cervix. It is almost unknown in nerve or muscle tissue. Cancer is an unchecked multiplication of abnormal cells in the body, which leads to the growth of an irregular mass of cancer cells called a tumour. The physical presence of the tumour may disrupt the structure and/or the functioning of normal tissues, causing death unless it is diagnosed at an early stage when treatment, which may involve surgery, chemical therapy or radiation therapy, is most effective.

The fundamental problem that is hindering the development of definitive cures for the various forms of cancer is the similarity between normal cells and cancer cells. For example, in the blood of normal people the number of white blood cells (leucocytes) is maintained at a fairly constant level, i.e. there is a balance between cell production and cell removal. In patients with leukaemia, leucocyte production is not controlled, and so very large numbers of leucocytes occur in the blood. The blood becomes 'choked' with leucocytes, and the normal functions of the red blood cell are seriously impaired. In many respects the normal cell and the cancer cell are identical, and so the problem relates to finding a cure that is sufficiently selective to destroy cancer cells but *not* normal cells. Many lines of research are being investigated, but from the standpoint of this book, the most interesting work has been concerned with a comparison of the fine biochemical characteristics of normal and cancer cells. It is hoped that such work may reveal a fundamental weakness in the cancer cell that could be exploited for therapeutic purposes.

A significant discovery in this line of research was made when research workers transferred serum from a healthy guinea pig into a mouse suffering from multiple leukaemias (see Chang, 1977). The mouse improved and survived, and it was deduced that some agent in the guinea pig serum was effective against the mouse cancer cells. A rapid search showed that the anti-leukaemia agent in the guinea pig serum was an enzyme called L-asparaginase. This enzyme hydrolyses the amino acid L-asparagine, which is one of the 20 or so naturally occurring amino acids found in proteins.

$$\text{L-asparagine} \xrightarrow{\text{L-asparaginase}} \text{L-aspartic acid} + NH_3$$

The anti-leukaemia activity of this enzyme is founded on a basic biochemical difference between normal cells and cancer cells. L-asparagine is an important amino acid that cells use in protein synthesis. It is generally regarded as non-essential because cells have the capacity to synthesize this amino acid. Essential amino acids such as leucine cannot be synthesized by cells and must be supplied in the diet. The terms essential and non-essential do not refer to the importance of the amino acid in protein synthesis, but to the need for the amino acid to be supplied in the diet or not.

In normal cells L-asparagine is non-essential because it can be synthesized, but in cancer cells it is essential and must be taken up into the cells from the blood. The addition of the L-asparaginase to the serum of the mouse caused a depletion of circulating L-asparagine, reducing the availability of this amino acid for the cancer cells, as illustrated in Fig. 4.5. The absence of this amino acid causes severe disruption of protein synthesis. Key proteins are not produced, and the cancer cell is unable to maintain its viability.

While this line of approach appears promising for cancer therapy, there are still a number of problems to be overcome before the method gains greater acceptability. In addition to the general problems associated with any enzyme therapy, preliminary clinical studies with human cancer patients have revealed severe side-effects. These have included nausea, vomiting, chills, diarrhoea, lethargy, serious liver and kidney disturbances. Such a wide spectrum of side-effects can probably be attributed to reduced synthesis of important proteins in the liver, pancreas and brain. It is likely that these side-effects are due to the inability of these tissues to react quickly to the rapidly depleting L-asparagine. While adequate supplies of the amino acid are circulating in the bloodstream, all cells will prefer to take-up the amino acid rather than synthesize it. Consequently, the metabolic machinery for producing L-asparagine will be more or less absent in normal cells, because it is not required. Faced with a sudden depletion of L-asparagine, the normal cell will try to initiate the synthesis of L-asparagine synthetase, which is responsible for the synthesis of the amino acid. However, the normal cell will be unable to synthesize the synthetase enzyme due to the disruption in protein synthesis caused by the depletion of the amino acid. This problem is probably not insurmountable, and may simply require that a very slow step-wise depletion of L-asparagine is undertaken to allow the normal cells time to produce the L-asparagine synthetase enzyme.

This useful discovery has stimulated further research work into the evaluation of enzyme-mediated nutrient depletion as a possible approach for cancer

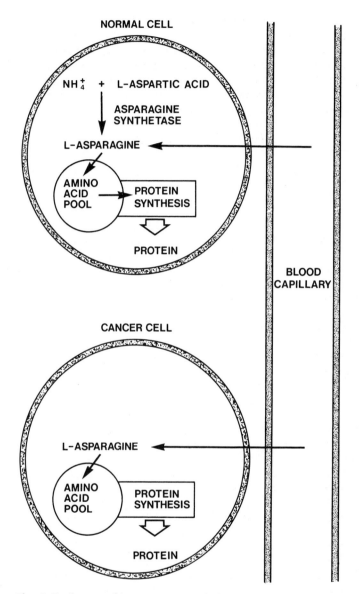

Fig. 4.5 Sources of L-asparagine supply for normal cells and asparagine synthetase deficient cancer cells.

therapy. It is hoped that other deficiencies may be discovered in other forms of cancer. Some progress has been made with the finding that other non-essential amino acids have become essential in certain cancer cell types. The search is being extended to consider vitamins, cofactors, sugars, lipids, nucleotides and other circulating metabolites, which may represent possible targets for depletion by enzymes.

Another interesting line of approach is the use of enzymes to stimulate the immune system to recognize that cancer cells are abnormal, and invoke a defensive response to remove the cancer cells. It has been known for a long time that the body defence systems are capable of selectively destroying cancer cells. It has been observed on a number of occasions that if cancer patients contract a serious life-threatening bacterial infection then, provided the patient survives, the body defences will destroy both the bacterial and the cancer cells. It would appear that when the full range of body defence systems is mobilized, then the great intensity of the response produces a mechanism capable of detecting and

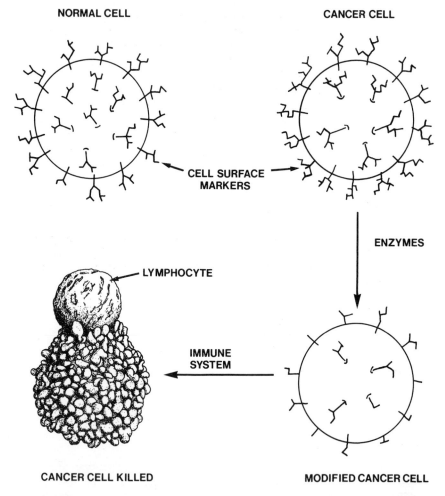

Fig. 4.6 Possible stimulation of lymphocyte-mediated destruction of cancer cells by alteration of cell surface recognition markers using enzymes.

destroying cancer cells. Research scientists are actively searching for a method to trigger the immune system into responding to the presence of cancer cells in the body.

One of the methods under investigation is concerned with the cell-surface markers possessed by all cells, which identifies them to the host defences as native and not foreign. These markers are sugar recognition chains and they form part of specific interaction sites on the cell surface. The pattern of the cell surface sugar chains determines the recognition code, and removal or alteration of the sugar chains causes the recognition mechanism to fail. The approach is illustrated in Fig. 4.6, and involves removing some cancer cells from the patient, then using enzymes to alter the cancer cell surface sugar recognition chains. The enzymes used are glycosidases such as neuraminidase (releases N-acetylneuraminic acid), galactosidase (releases galactose), mannosidase (releases mannose) and fucosidase (releases fucose). The altered cells are then returned to the patient in the hope that the cancer cells may be recognized as foreign and invoke an immune response. However, although the returned (altered) cancer cells are rapidly destroyed, no immune response is generated by this procedure. The method may have a better chance of success when more details are known about the cell surface determinants responsible for specifying likeness between cancer cells and normal cells, and the surface determinants on bacteria that readily stimulate an immune reaction. Recent research suggests that it may be more profitable to enzymically add more sugars onto the surface of the cancer cells to create a more foreign and, therefore, more antigenic surface.

4.4   Protein degradation

The degradation of protein macromolecules is catalysed by a group of enzymes called proteases. These enzymes have a significant and increasing application in many areas of medicine, for a wide range of purposes connected with removal or degradation of protein.

$$\text{protein} \xrightarrow{\text{protease}} \text{peptides + amino acids}$$

Many different protease enzymes are available – a short list is given in Table 4.2. The selection of a particular protease for a given purpose depends largely

Table 4.2   Some proteolytic enzymes used in medicine.

Enzyme	Source
Trypsin	Mammalian pancreas
Chymotrypsin	Mammalian pancreas
Bromelains	Pineapple juice
Papain	*Carica papaya* plant latex
Pepsin	Mammalian stomach
Plasmin	Blood
Urokinase	Urine
Streptokinase	*Streptococcus* bacteria

on the protein to be degraded, and on the extent of degradation required. The enzymes have different specificities, which restricts the range of proteins that an enzyme can degrade. For example, trypsin has a fairly narrow specificity, and in most cases it has only a weak degradative action on native proteins. Pepsin, however, has a wider specificity and catalyses the hydrolysis of almost all native proteins. In medical applications, it is normally desirable to use the slow-acting enzymes of narrow specificity. Thus minimizing harmful degradation of tissue protein, which may be surrounding the target protein. Often it is useful to use a combination of several enzymes with different specificities to obtain the desired level of protein degradation.

The use of proteases to assist the natural healing of wounds dates back to early military and naval surgeons. The proteases were obtained from the mouth parts of live maggots, and used to clean suppurating wounds sustained by soldiers and sailors in battle. In modern times, enzymes are supplied in bottles, but the application is still the same. Treatment of serious skin damage usually necessitates the removal of damaged necrotic tissue or hard scab material, in order to clean and sterilize the wound. The formation of a clot and then a scab protects a wound from further invasion by bacteria, but also traps bacteria already present in the wound. The trapped bacteria are able to multiply and delay the healing process by obstructing the renewal of skin epidermis. An exudation of pus from a wound is clear evidence of bacterial presence in a wound. In serious cases where a large clot has formed and the wound is not healing normally, the scab may be treated with proteases such as trypsin and chymotrypsin. The enzymes degrade the fibrin-based scab, which returns to a fluid consistency that is easily removed before the application of antiseptics to sterilize the wound. A variety of special bandages and compresses, which are impregnated with protease enzymes and antiseptics have been developed, and found to be useful in speeding-up the healing of damaged skin.

In the treatment of severe skin burns it is often necessary to graft healthy skin onto the damaged area. Skin grafts provide protection and enhance the repair of damaged skin. Before attempting skin grafts it is essential to remove debris, damaged or necrotic tissue, and provide a clear foundation surface for the skin graft. The damaged skin may be cleaned using surgical techniques or by the application of protease enzymes, and frequently a combination of both is used depending on the type of skin damage.

Protease enzymes have an important use in treating thrombosis, which is the blockage of blood vessels by fibrin-based clots, and is responsible for almost half of the deaths in developed countries such as Britain. In particular, coronary heart disease is caused by blockage of blood vessels serving the heart, pulmonary embolism by blockage of blood vessels serving the lungs and a stroke is caused by blockage of blood vessels serving the brain. Present therapy employs anti-coagulants such as heparin and warfarin that are effective but slow-acting, and protease enzymes to promote the natural lysis of a fibrin clot.

The blood contains plasmin, a natural fibrinolytic protease enzyme, in the form of an inactive precursor called plasminogen. The enzyme thrombolytic therapy involves the catalytic conversion of plasminogen to active plasmin in the blood by an added protease enzyme called a plasminogen activator (PA), as indicated on the left-hand side of Fig. 4.4 (p. 54). Several protease enzymes are capable of catalysing the conversion of inactive plasminogen to active plasmin, thus facilitating dissolution of fibrin clots. One important plasminogen activator is the bacterial protease streptokinase (s-PA), which is widely used for eliminating blood clots from the lungs. Another useful PA is urokinase (u-PA) from urine, and this protease has been tested for possible use in coronary heart disease. However, both of these enzymes have drawbacks, for example streptokinase causes immunological complications, and urokinase is very expensive to isolate from urine. Also, streptokinase lacks specificity as it activates plasminogen molecules throughout the blood, which may create a serious risk of haemorrhage in other parts of the body. Urokinase is more specific, and only activates plasminogen at the site of the clot formation.

Natural plasminogen activators are found in blood, tears and other body fluids as well as in the cells of many tissues, including cancer cells. Recent research has isolated a tissue plasminogen activator (t-PA) from cultures of human cancer cells, and preliminary studies with this PA appear very promising. One hospital study has made use of t-PA in emergency cases such as patients in the throes of having a heart attack. The results of this trial showed that in almost all cases, the heart attack dispersed within 30–60 minutes after injection of the t-PA. A considerable advantage of the t-PA is specificity, it requires the presence of fibrin so only activates plasminogen at the site of the clot, thus greatly reducing the risk of haemorrhage. Unfortunately, t-PA is produced naturally in very small quantities, and it will be necessary to produce the enzyme by genetic engineering methods before t-PA will be readily available. It is likely that wider availability of this enzyme would dramatically reduce the number of deaths caused by heart attack (see Gronow and Bliem, 1983).

It should be recognized that in addition to removal of unwanted protein, proteases have a great potential for destruction of healthy tissue. Consequently, proteases are normally restricted to specific areas in the body where they may operate in a controlled manner. In addition to control by compartmentation, there are a number of potent protease inhibitors that circulate in the blood (Table 4.3). These inhibitors are themselves small proteins which serve to regulate the natural activity of blood proteases, and protect the blood compartment from the destructive effects of 'wild' proteases that 'escape' into the bloodstream. However, in some diseases an imbalance in the ratio of protease-protease inhibitor can occur thus allowing local protein degradation to destroy body tissue. Also some types of bacteria secrete protease enzymes that degrade

Table 4.3   Some proteins found in human blood plasma that inhibit protease enzymes.

Inhibitor	Proteases inhibited
Alpha$_1$-Antitrypsin	Elastase, trypsin
Alpha$_1$-Antichymotrypsin	Cathepsin G, chymotrypsin
Alpha$_2$-Macroglobulin	Wide range of proteases
Alpha$_2$-Antiplasmin	Plasmin
Antithrombin	Thrombin

body tissue to facilitate the invasion of bacterial cells and the consolidation of bacterial infection. Much research effort is being devoted to the identification, extraction and evaluation of protease inhibitors that may be of value in medical therapy.

In emphysema the disruption of lung tissue involves the participation of a protease called elastase which is secreted by neutrophils (a type of white blood cell). In this disease, the tiny air sacs of the lungs become over-inflated, their walls become distended and lose vital elasticity. The protease attacks the elastic protein (elastin) of lung tissue causing a progressive reduction in elasticity, and an increase in breathlessness as the efficiency of the lung tissue is impaired. In the absence of disease or dysfunction, elastase is inhibited by alpha$_1$-antitrypsin which protects the lung from the destructive activity of the protease. In some diseases, for example early-onset of pulmonary emphysema, patients are deficient in or have a defective alpha$_1$-antitrypsin inhibitor. Emphysema normally accompanies other lung diseases such as chronic bronchitis and lung cancer. It has been recognized for some time that cigarette smoking aggravates emphysema as cigarette smoke contains chemicals that oxidize vital amino acids in the active centre of the alpha$_1$-antitrypsin inhibitor (e.g. methionine is oxidized to methionine sulphoxide) and render the inhibitor inactive. Therefore, cigarette smoke is likely to enhance emphysema, and since the flow of blood is slowed down in disrupted lung tissue, an added strain is put upon the heart which may contribute to coronary heart disease.

A number of protease inhibitors are currently being used or evaluated for medical therapy, for example alpha$_1$-antitrypsin inhibitor is used in the treatment of pancreatitis, where proteases 'escape' into the blood compartment (see pp. 32–3), and it is likely that greater availability of protease inhibitors may help to improve treatments in a wide range of diseases where protein degradation is involved either directly or indirectly. For further discussion on the molecular physiology and pathology of alpha$_1$-antitrypsin see Boswell and Bathurst (1985).

4.5   Removal of toxic compounds

Potentially harmful compounds can occur in the body as a result of a variety of circumstances, ranging from everyday consumption of non-nutritive substances to deliberate drug abuse and metabolic disease. It is a fact of life that humans are constantly exposed to a wide range of potentially toxic materials. Many of these compounds are inherently present in some foodstuffs or are generated as a result of food processing, for example nitrosamines are formed when added nitrites interact with amines in food products. Others are consumed in the form of medicines (e.g. aspirin, paracetamol, valium), or in beverages (e.g. caffeine, tannins, ethanol). In general, the liver and kidney work together to eliminate such compounds, and thus prevent a build-up of higher concentrations that may become toxic.

From a medical standpoint, the necessity for removing harmful compounds from the body arises when a large and potentially harmful amount of material is taken into the body, or when a failure occurs in the normal functioning of the liver and/or kidney such that compounds accumulate in the body to toxic levels. Liver cells employ a number of sophisticated enzyme systems to metabolize and transform harmful compounds to forms more suitable for disposal by the kidney. It follows that enzymes are ideally suited for removing these compounds from the body when the removal systems are overwhelmed, for example, in drug overdose or organ failure. Although most of the harmful compounds tend to accumulate in the blood, direct injection of soluble enzymes into the bloodstream is rarely undertaken. However, in emergency situations where rapid action is required to deal with a particular problem, for example, mushroom poisoning or snake bite toxin, then direct injection of specific enzymes may be required to avoid a fatality.

For longer-term therapy involving specific enzymes to detoxify compounds, it is necessary to use immobilization techniques to protect the enzymes from body defence mechanisms. As indicated earlier, this can be achieved by immobilizing enzymes in a semi-permeable membrane capsule (see Fig. 4.2, p. 49). With such systems, the toxic compound permeates into the capsule, is transformed by the enzyme, and the harmless product(s) diffuse out of the capsule. Large molecules such as antibodies and body defence cells cannot enter the microcapsule, thus avoiding immunological reactions. This type of enzyme system has great potential for improving the therapy currently available to patients with acute kidney failure.

Normally the kidney excretes on average one millilitre of urine per minute, containing about 0.025 g of urea (principal breakdown product of proteins). In patients with kidney failure, urea is not excreted and accumulates in the blood. The body can tolerate fairly high concentrations of urea, but if it is not removed then the steady accumulation soon reaches a harmful level. Current therapy involves removing urea from the blood using a dialysis machine – a basic scheme for dialysis is shown in Fig. 4.7a. Patients normally attend hospital two or three times per week to spend time on a dialysis machine. The machine works on the simple diffusion principle that if one liquid (the blood) containing a high concentration of urea is separated, using a semi-permeable membrane, from another liquid (dialysis fluid) containing no urea, then the natural process of equilibration will cause the urea to diffuse through the membrane from the blood into the dialysis fluid. The dialysis machine is effective for removing urea from the blood, but has the disadvantage that it is expensive to supervise and operate. Also the machine requires large volumes of dialysis fluid, and produces large volumes of dialysate, which contains urea, and therefore cannot be re-used.

An alternative system has been developed employing the enzyme urease to eliminate urea from the dialysate, and thereby allow the dialysate to be recycled. This process substantially reduces the volume of dialysis fluid required, and therefore a smaller, more mobile dialysis machine can be used (Fig. 4.7b). Urease is immobilized by entrapment in semi-permeable microcapsules, and these are placed beside a compartment of activated carbon in a reactor column. The blood is routed from the patient to the compact dialyser, where urea and other metabolites diffuse into the dialysate. Then the blood is

(a)

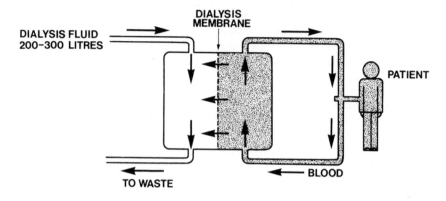

(b)

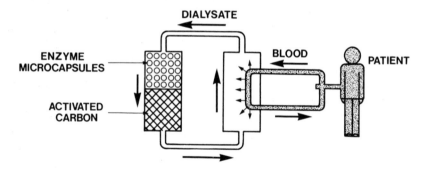

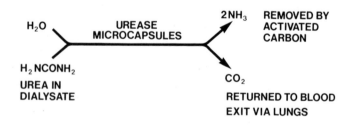

Fig. 4.7 Removal of urea from blood by **(a)** conventional dialysis, and **(b)** by hydrolysis of urea using immobilized urease.

returned to the patient, and the dialysate is pumped through the reactor column. Urea is hydrolyzed to ammonium ions and bicarbonate ions by urease located in the reactor column. Ammonium ions and some other waste metabolites are retained by the activated carbon. The bicarbonate ions stay in the dialysate and are returned to the blood as the dialysate is re-cycled, eventually being eliminated by the lungs as CO_2. This system has been evaluated, and has proved very effective for removing urea. Also, the urease procedure is more economical to operate than the conventional machine. Future developments could dispense with the need for a dialysis machine if suitably prepared microcapsules are injected into the body, or a small disposable reactor column is connected to the body. However, several aspects have still to be evaluated before such new developments can be considered as suitable replacements for the dialysis methods.

A considerably more difficult problem is encompassed in the prospect for developing a therapy for patients with liver failure. The liver has many varied functions including formation of bile, storage and metabolism of carbohydrate, metabolism of hormones, drugs and other compounds, destruction of worn-out red blood cells and fat metabolism. With present day technology, it is not conceivable that a constructed system could be developed to take over all the vital functions of the liver. At best, it might be possible to provide limited support in cases of partial liver failure. A hepatic-assist device is being investigated that uses isolated microsomes, which contain drug metabolizing enzymes, for possible application in supporting the liver in cases of drug overdose or partial liver damage due to disease. Patients with extensive liver disease or total liver failure require a liver transplant, and it is likely that for a long time, this will remain the only method of treatment.

4.6 Prevention of dental caries

Dental caries is the loss of tooth material called hydroxyapatite by acid dissolution, producing cavities in the teeth and toothache as the cavity exposes a nerve ending. It is an extensive problem, which is reflected in the fact that a high proportion of older adults have lost most or all of their natural teeth as a consequence of tooth decay and gum disease. Also, it is probably the most costly of the diseases afflicting humans. It has been estimated that in Britain over 2.5 million working days are lost annually, and over £500 million is spent each year treating this disease and peridontal disease.

In the absence of effective and regular oral hygiene, the teeth are soon covered by a yellowish film of material called dental plaque. This is composed of bacteria, proteins, food particles, cellular debris, carbohydrate and inorganic salts. It has been found that without oral hygiene, a covering of plaque up to 0.2 g (wet weight) and containing 10^{10} bacterial cells may accumulate on the teeth. Although many different types of bacteria may be present in plaque, tooth decay is associated with anaerobic bacteria, and in particular *Streptococcus mutans*. This is because anaerobic bacteria produce acids, for example lactic acid as end products of carbohydrate metabolism, and there is a very strong correlation between carbohydrate intake, acid production, and the incidence of dental caries. For example, sucrose is used by this organism to synthesize a glucose-based polysaccharide called mutan. The mutan helps to cement the plaque onto the tooth enamel, and is used as a source of sugars to sustain the bacteria during periods of no food intake such as between meals and overnight fasting.

Tooth decay can be prevented by the use of simple, but effective preventive measures. These include physical removal of dental plaque by brushing the teeth, coating the teeth with an acid resistant plastic polymer, and addition of fluoride to strengthen the hydroxyapatite resistance to acid dissolution and inhibit bacterial metabolism. An enzyme-based method has been proposed that makes use of the natural anti-bacterial enzyme found in saliva, called lactoperoxidase. This enzyme readily binds to bacteria and catalyses the production of hypothiocyanite ions ($OSCN^-$) from thiocyanate (SCN^-) and hydrogen peroxide.

$$H_2O_2 + SCN^- \xrightarrow{\text{lactoperoxidase}} OSCN^- + H_2O$$

The $OSCN^-$ ions are potent anti-bacterial agents that inhibit bacterial enzymes by oxidizing sulphydryl (SH) groups in certain enzymes. However, although SCN^- and lactoperoxidase are both present in saliva, the concentration of H_2O_2 in saliva is very low. Thus the natural anti-bacterial lactoperoxidase is normally constrained by the absence of H_2O_2, and therefore provides only limited protection against dental caries. In order to increase the amount of H_2O_2 available to the lactoperoxidase, the enzymes glucose oxidase and glucoamylase have been incorporated into a new toothpaste formulation. The glucoamylase produces glucose from starch in the diet, and the glucose oxidase produces H_2O_2 from the oxidation of glucose.

$$\text{starch} \xrightarrow{\text{glucoamylase}} \text{glucose} \xrightarrow{\text{glucose oxidase}} \text{gluconic acid} + H_2O_2$$

Tests have shown that plaque formation is reduced in subjects using the new formulation, and due to the success of the tests an enzyme toothpaste is now available in several countries under the trade name of 'Zendium'. This product contains 0.3% w/w glucose oxidase, 1.2% w/w glucoamylase and 0.02% w/w potassium thiocyanate as the key ingredients. However, it was noted during some of the trials that the product was less effective when used by some of the more meticulous subjects who, after brushing their teeth, habitually rinse out their mouth with water three or four times.

5

Industrial Applications

5.1 Introduction

The specialized catalytic capabilities of enzymes have been used extensively for thousands of years in traditional industries such as wine, beer, bread and cheese making. No doubt this usage will continue for a long time to come. With the greater appreciation of the role of enzymes in these industries, the potential for using enzymes as catalysts in other areas such as analysis, medicine, chemical synthesis and conversions, has been widely promoted. The scope for using enzymes as industrial catalysts is indicated by the wide range of reaction types that can be catalysed by enzymes. These include oxidation/reduction, inter- and intramolecular transfer of a variety of chemical groups, hydrolysis, cleavage of covalent bonds, isomerization and addition of chemical groups across double bonds; so almost all organic and many inorganic reactions can be catalysed by one or more enzymes. However, many of these reactions can also be catalysed by a variety of non-biological catalysts. The importance of these agents is illustrated by the fact that 70% of all the current chemical process industry involves catalysis, and the majority of it uses non-biological catalysts.

With two or three exceptions, the existing chemical processing industry has not made very much use of enzyme catalysts. The principal stumbling block is the economic balance between old and new procedures. Traditional systems may be less efficient and more expensive to operate, but they are already equipped with plant and machinery, and they are run by trained people who are familiar with the components and therefore solve operational problems quickly. A replacement process normally involves a major financial investment to re-equip with new machinery and retrain personnel to operate the process. Such an investment would only be considered viable if the new process was likely to produce a significant financial return within a reasonable time scale, so it is likely that enzyme catalysts will continue to make slow progress in existing industry, but this will be offset by rapid progress and development in new industry. In medical and analytical fields, the level of capital outlay required is much reduced, and so the principal considerations relate to the effectiveness of the enzyme and any problems associated with its application. The aim of this chapter is to illustrate the considerable potential of enzymes for application in a wide array of large and small scale processes in industry, and it is hoped that this will be best achieved by describing the involvement of various enzymes in a number of different situations.

5.2 Sugar production

The sugar industry is a major consumer of enzymes, which are used in the hydrolysis of starch to produce partially hydrolysed starch syrups (called modified starch or dextrins), glucose (also called dextrose) and fructose for application in food processing and preparation. Starch is the major reserve carbohydrate in all higher plants, and it is composed of both linear (amylose) and branched polymers (amylopectin) of glucose (Fig. 5.2). It is produced

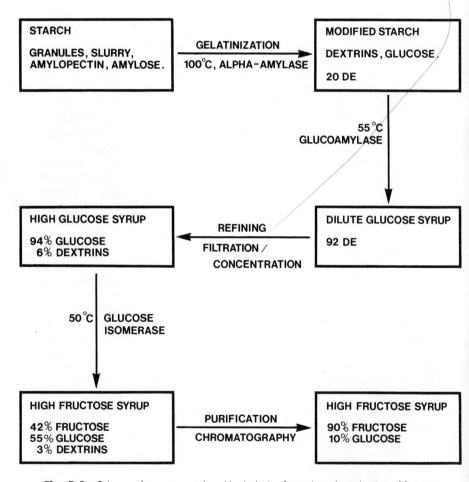

Fig. 5.1 Scheme of enzyme catalysed hydrolysis of starch and production of fructose.

commercially from seeds, tubers and roots, and the major sources of starch in the Western Hemisphere are maize (corn) and potato.

Early procedures for the hydrolysis of starch employed acid. However, this method was not very economical giving a fairly low yield of glucose, and the product contained undesirable coloured compounds that had to be removed by chromatography. Enzymes have replaced the original acid procedure, and brought improvements in the efficiency of the hydrolysis, and in the quality of the final product (see Godfrey and Reichelt, 1983). An outline of the sequential steps in the enzymic hydrolysis of starch is given in Fig. 5.1.

Before starch can be hydrolysed it must first be gelatinized (put into solution). This is accomplished by heating the insoluble starch slurry at 100°C to swell and rupture the starch granules. Alpha-amylase from *Bacillus licheniformis* is added during the gelatinization to prevent the starch paste from becoming too thick, and to begin the enzymic hydrolysis of the glucose polymers. The enzyme from this particular organism displays high thermal stability (catalytically active at 115°C), and it therefore suitable for use at the high temperatures required for gelatinization. Alpha-amylase is an endo-attacking enzyme that hydrolyses the 1,4 alpha-linkages between the glucose residues in the polysaccharide chains (Fig. 5.2a). This limited hydrolysis produces a variety of lower molecular weight saccharide units, for example from decasaccharides to disaccharides (collectively called dextrins), but only a minimal amount of monosaccharide glucose. The extent of the degradation can be measured and expressed in terms of dextrose equivalent (DE). Pure glucose has a DE value of 100, and unhydrolysed starch has a DE value of zero.

A range of modified starch products are produced, varying in the extent of hydrolysis, which have extensive application in the food preparation industry. Low DE starch syrups are used in frozen mousses as they slow the melt down of such products when they are returned to room temperature. They are also used as thickening agents/stabilizers in trifles, ice-creams, cake mixes, milk shakes, sauces etc. High DE starch syrups are used to stabilize the colour in jellies and jam, and to impart body and smoothness into many products.

In the next stage, conversion of the dextrins to glucose is accomplished using glucoamylase from *Aspergillus* fungus. This is an exo-attacking enzyme that consecutively releases glucose residues from the dextrin fragments (Fig. 5.2b). The conventional process uses soluble enzyme added to a stirred dextrin solution in large tanks, and it normally takes 50 to 60 hours to produce a glucose syrup with a DE value of 92. Some interesting pilot plant studies have investigated the possibility of using an immobilized glucoamylase (see Mosbach, 1976). In this system the immobilized enzyme was packed into a reactor column, and the required reaction time was effectively reduced down to two hours. However, the high operating temperature (50°C) used in the soluble enzyme process is less suitable for the immobilized enzyme as the heat promotes release of the enzyme from its support material.

After some refining and concentration, the resulting glucose syrup can be spray-dried or crystallized to form a solid glucose product. Glucose is used extensively as an additive in a wide range of produce, from soft drinks, health drinks, packed cake mixes, confectioneries, whipped cream, frozen desserts to

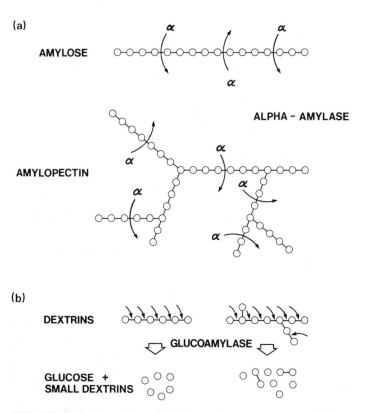

Fig. 5.2   A representation of **(a)** the endo-enzyme activity of alpha-amylase, and **(b)** the exo-enzyme activity of glucoamylase.

vinegar, pickled onions and chutney. It is not as sweet as sucrose (Table 5.1) so it has limited use in very sweet products; rather it is used to contribute a background sweetness. In fermented products such as vinegar it serves as a substrate for limited fermentation in the bottle.

The final stage in the process, which has been developed in the last 12 years, is the conversion of glucose to its stereoisomer fructose (fruit sugar). This sugar is sweeter than sucrose (Table 5.1), and is therefore suitable for application in the very large market that exists for principal sweeteners. (See Hough and Emsley (1986) for further discussion on the development of sweeteners.) There is no chemical process available for the highly specific conversion of glucose to fructose, so the reaction is accomplished using the enzyme glucose isomerase.

$$\text{glucose} \xrightarrow{\text{glucose isomerase}} \text{fructose}$$

The substrate for the reaction is the high glucose syrup (HGS) containing 94% glucose and 6% dextrins, and the enzyme catalyses the isomerization of

Table 5.1 Comparative sweetness of various carbohydrate products.

Carbohydrate	Sweetness (relative to sucrose)
Sucrose	1.00
Fructose	1.50
Glucose	0.75
Lactose	0.16
Starch syrup (DE 35)	0.26
Starch syrup (DE 60)	0.45
High fructose syrup (42% fructose)	1.00

glucose to fructose until the reaction reaches equilibrium. At this stage the product consists of 42% fructose, 55% glucose and 3% dextrins and is called high fructose syrup (HFS). Glucose isomerase is an intracellular enzyme found in various micro-organisms, and normally the enzyme is used *in situ*. Evaluation studies have demonstrated that there is no advantage in productivity or enzyme stability in using an isolated/purified enzyme. Most companies involved in this work use a dead microbial cell mass in which the crude microbial cell mass is cross-linked with a fixative (glutaraldehyde). The cross-linked cell mass can be made into pellets by extrusion, and the pellets packed into a reactor column. Thus, the enzyme is effectively immobilized inside the microbial cell mass. In this state the enzyme is both stable and catalytically efficient.

Lastly, further refining involving chromatographic separation produces a HFS consisting of 90% fructose and 10% glucose. The total production of HFS in Europe and the USA is estimated at three million tonnes per annum, and much of this will be used in the replacement of sucrose as the principal sweetener in soft drinks, confectioneries, sweet foods etc. Several years ago the giant Coca-Cola company announced that it was replacing sucrose with HFS in its soft drinks products, so it would seem that there is a large and growing future for this aspect of enzyme technology. Although HFS is manufactured in Europe, successful application of the product has been limited in EEC countries because of strong political influences seeking to protect sugar cane/beet farmers from an anticipated decline in the use of sucrose.

5.3 Enzymes in detergent products

Most of the current detergents designed for home laundry purposes are synthetic soap powders derived from petroleum fractions. They normally contain a variety of additives for improving the effectiveness of the basic detergent, and giving the impression of greater brightness in the washed material. Over the last 20 years the development of 'biological washing powders' has seen a significant expansion in the addition of enzymes to washing detergents for use in the home.

The 'dirt' in clothes normally consists of various inorganic compounds together with biological material such as protein, fat and carbohydrate. Dirt may come from many sources, but much of the biological material found on clothes arises from food, drink, dead skin cells, blood or body secretions and excretions such as perspiration oils, and urine. Protein stains are notoriously difficult to remove from cloth fabric, especially if the protein has dried out on the material. As the protein loses water, it denatures and the polypeptide chains form cross-linkages with each other and with the fibres of the cloth. This produces an insoluble precipitate that is firmly enmeshed in the fibres, and very resistant to the normal lifting/dispersing activity of a basic detergent. Traditionally, such stains were removed by using a very hot wash (85°–95°C) to fragment and dislodge the protein precipitates. However, such high temperatures are expensive in terms of energy usage, and damaging to many types of cloth causing shrinkage and loss of coloured dyes into the wash.

The addition of protease enzymes to washing detergents has created an expanding market in 'biological washing powders'. The proteases effectively break down the polypeptide chains of the protein into soluble peptides and amino acids that can be removed in the wash. The particular enzymes normally used are bacterial proteases from organisms such as *Bacillus subtilis* and *Bacillus licheniformis*. The proteases from these organisms are stable in the alkaline conditions required for detergent activity, and are compatible with the other additives such as sequestering agents, builders, bleaches, conditioners and optical brighteners. The bacterial proteases also retain their catalytic activity at temperatures up to 65°C for about an hour or two. Most products contain sufficient quantity of enzyme to provide a final concentration in the wash of about one milligram of enzyme per litre of solution.

The enzyme washing powders were quickly adopted by the detergent industry, and early experiences with these products highlighted the need for proper evaluation of the hazards in working with enzymes. During the manufacture, production workers developed allergies to enzyme dust, and some sensitive customers complained of skin reactions. This problem was rapidly identified and remedied by producing dust-free products. The enzymes are temporarily immobilized by encapsulation in special granules that are mechanically flexible and elastic, giving resistance to crushing and dust formation during production. In contact with water, the granules swell and rupture, releasing the enzyme into the wash solution.

In the future it is likely that the trend will continue towards lower washing temperatures as energy costs increase, and this will necessitate a re-examination of product formulations to maintain the performance of washing detergents. Research investigations are evaluating the possibility of including lipases for fat dispersal and amylases for starch breakdown in new formulations. New enzymes capable of breaking down hydrocarbon waxes and oils could have use in several engineering industries. Perhaps the most radical possibility is the use of enzymes to modify proteins and triacylglycerol lipids to form a new range of efficient biological surfactants, to replace the traditional sulphonate-based surfactant used in current detergent formulations.

5.4   Production of semi-synthetic penicillins

The penicillins are a group of comparatively small compounds with important anti-bacterial activity. The group is composed of both naturally produced and semi-synthetic penicillins, and as a group they are used to check the growth of a range of susceptible bacteria. In particular, they are effective against the micro-organisms responsible for diseases such as tetanus, diphtheria, syphilis, gonorrhoea and skin boils. Structurally, the group of molecules have a common characteristic fused-ring structure, and differ only in the nature of the side chain attached to the beta-lactam ring (Fig. 5.3).

Of the naturally produced penicillin molecules, only penicillin G has strong anti-bacterial activity, and consequently only penicillin G is produced in large quantities by industrial fermentation using *Penicillium* species. However, the usefulness of penicillin G in clinical therapy is steadily declining as increasing numbers of bacteria acquire resistance to the antibiotic. The resistant strains are able to produce a beta-lactamase enzyme that destroys the beta-lactam ring

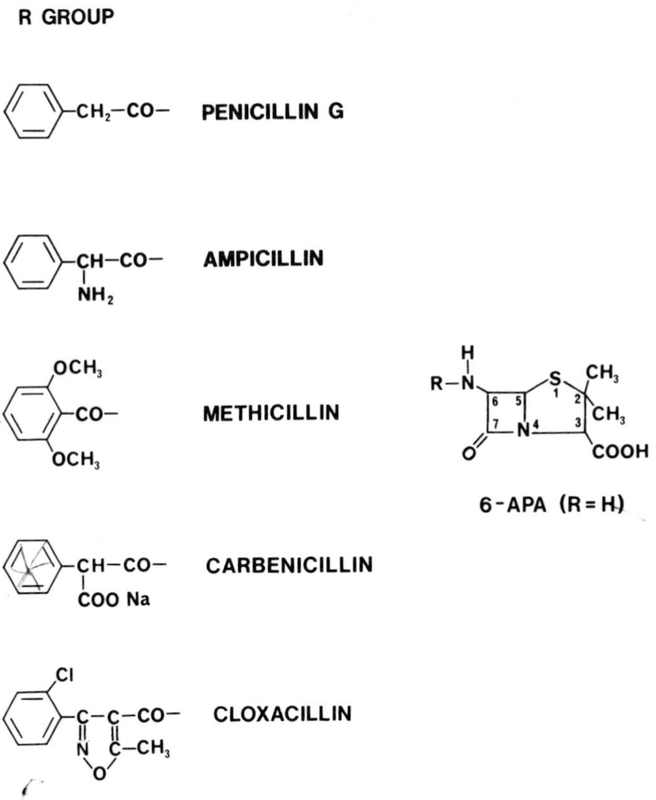

R GROUP

Fig. 5.3 Structural comparison between some semi-synthetic penicillins and naturally produced penicillin G.

structure, and renders the antibiotic inactive. A positive awareness of this growing problem has stimulated much research to find new antibiotics, and although a great many synthetic and natural compounds have been tested, only very few have been considered suitable for clinical use. One approach which has proved successful with the penicillins involves the simple logical concept of turning the problem on its head. If bacteria are resistant to penicillin because of beta-lactamase, simply modify the penicillin and make it resistant to the activity of the beta-lactamase. The successful development of this idea has provided a range of semi-synthetic penicillins having distinct advantages over the naturally produced parent molecules.

The modification is accomplished by first removing the side-chain from the naturally produced penicillin G to provide the basic nucleus of the antibiotic called 6-aminopenicillanic acid (6-APA). It is then comparatively simple to join another side chain substituent onto the 6-APA to produce a semi-synthetic structural analogue composed of a naturally derived nucleus and a synthetic side-chain. Prior to 1970, hydrolysis of the side-chain was achieved by a chemical process involving organic solvents and low temperatures ($-50°C$). An acceptable enzyme-based alternative was subsequently developed in the late 1960s using penicillin acylase.

$$\text{penicillin G} \quad \xrightarrow[\text{H}_2\text{O, pH 8, 37°C}]{\text{penicillin acylase}} \quad \begin{array}{l} \text{6-aminopenicillanic acid (6-APA)} + \\ \text{phenylacetic acid (side-chain)} \end{array}$$

The acylase enzyme is widely distributed in bacteria and fungi, and a common source of enzyme for the hydrolysis of penicillin G is from *Escherichia coli*. Because of the release of acid during the reaction, it is necessary to stabilize the pH of the reactor by continuous addition of alkali. Regular monitoring of alkali addition also serves as a useful check on the progress of the reaction. To facilitate the alkali addition, the reaction is normally performed in a large stirred tank batch operation. When the reaction is completed, the enzyme is withdrawn and the phenylacetic acid is removed using solvent. The 6-APA is collected by adjusting the pH to pH 4.3, the isoelectric point of 6-APA, at which the 6-APA precipitates out as a fine white crystalline material. An advantage of the *E. coli* enzyme is that it readily catalyses the reverse reaction. By altering the reaction conditions (e.g. to pH 5.0) the same enzyme can be used in the synthesis of an analogue penicillin using 6-APA and a new side chain.

$$\text{phenyl glycine} + \text{6-APA} \quad \xrightarrow[\text{penicillin acylase}]{\text{37°C pH 5.0}} \quad \text{ampicillin}$$

A number of preparations of immobilized penicillin acylase have been evaluated, and a few have been adopted in industrial scale processes. Early systems used intact *E. coli* cells immobilized by ionic attraction onto DEAE-cellulose. In other systems, purified enzyme from *E. coli* has been entrapped in cellulose triacetate fibres, and covalently bound to CNBr-activated Sephadex. The largest producer of semi-synthetic penicillins in Britain is Beecham, and in their system the enzyme is immobilized covalently to a polymethacrylate resin using glutaraldehyde. In almost all cases, the 6-APA is obtained with high purity ($\sim$ 98%) and high yield ($\sim$ 90%). The immobilized enzyme preparations are stable and enable good productivities to be achieved, for example 1500 kg 6-APA per kg of enzyme preparation. The current world production is probably about 12 000 tonnes per year, and about 70% of this is produced by the use of enzymes.

Several of the semi-synthetic penicillins are shown in Fig. 5.3. The modified penicillins have three distinct advantages over the natural penicillin G. Firstly, some of them (ampicillin and cloxacillin) are acid stable, and are therefore, suitable for oral intake. Secondly, methicillin and cloxacillin are resistant to beta-lactamase activity. Lastly, the group as a whole have significantly increased the range of susceptible bacteria to include many bacteria that are not inhibited by penicillin G.

5.5 Hydrolysis of lactose

Lactose (milk sugar) is a disaccharide sugar that occurs naturally in the milk of mammals, and it is the principal component of whey (a by-product of cheese production). It is not very sweet, has low solubility, and is not absorbed by the digestive system. However, upon hydrolysis by the enzyme beta-galactosidase (lactase), it is broken down to its constituent monosaccharides glucose and galactose.

$$\text{lactose} + H_2O \xrightarrow{\text{beta-galactosidase}} \text{glucose} + \text{galactose}$$

There is significant industrial interest in the hydrolysis of lactose both in milk and in whey, and the availability of beta-galactosidase has opened-up several interesting developments in the dairy industry.

In particular, there is a strong desire to hydrolyse lactose in milk to overcome the problem of lactose intolerance, which occurs in large sections of the poorer populations of the world. Lactose intolerance arises from a deficiency of beta-galactosidase in intestinal mucosa so lactose accumulates in the intestine, and serves as a carbohydrate source for intestinal microflora. As a result, lactic acid and CO_2 are formed, causing gastrointestinal irritation, abdominal pain and diarrhoea. Lactose intolerance is common among the adult populations of Asia, Africa, Latin America and the Middle East. Consequently, there is a need for the reduction of lactose in milk so that it and other dairy products can be made more readily acceptable to this potentially very large market. In addition, lactose-reduced milk will improve the preparation of concentrated milk products, where the low solubility of lactose often leads to the formation of crystals which cause the 'sandy' texture found in ice cream after prolonged cold storage.

Preliminary investigations on a pilot plant scale in Italy over 12 years ago demonstrated the successful hydrolysis of lactose in skimmed milk, and led to the development of a full scale industrial plant capable of treating ten tonnes of skimmed milk per day. An outline scheme of the pilot plant operation is shown in Fig. 5.4. The enzyme used was beta-galactosidase from the yeast *Saccharomyces lactis*, and it was immobilized by entrapment in cellulose triacetate fibres.

The process was operated on a batch mode, treating 200 litres per batch. After sterilization, the batch of skimmed milk was continuously circulated through the column reactor at a rate of seven litres per minute for about 20 hours. The milk was then allowed to accumulate in the storage tank before packaging. The process was operated at 7°C in order to reduce the possibility of bacterial spoilage. Studies revealed that the economic feasibility of the process depended on the level of enzyme activity, because this factor determined the ratio of milk volume to immobilized enzyme volume. The method of enzyme immobilization, entrapment in fibres, was selected because it enables the immobilization of high concentrations of enzyme in comparatively small volumes. Analysis of the reaction process demonstrated that maximum catalytic efficiency could be achieved using 500 g of fibres containing 1500 units of enzyme activity per gram of fibre. Higher levels of enzyme activity were tested, but found to cause a reduction in the diffusion of lactose into the fibres, resulting in reduced catalytic efficiency. The immobilized enzyme preparation proved to be very stable, with the loss of enzyme activity being less than 10% after processing 10 000 litres in 50 consecutive batches.

A major new area of development in enzyme technology is concerned with the processing and utilization of cheese whey. Whey is a liquid by-product of the cheese-making industry that arises from the separation of the casein protein and butter fat as curd from the milk. On average, whey contains about 10% dry material of which 8.5% is lactose, 0.9% is protein and 0.6% is salts. In order to produce 500 g of cheese it is necessary to use about 5 litres of milk, and this produces about 4.5 litres of whey. On a national scale, it has been estimated that the British cheese-making industry produces over 1000 million litres of whey each year.

For many years whey was regarded as a waste product to be disposed of in the sewerage system, generating a serious pollution problem because whey imposes a high biochemical oxygen demand in receiving waters. In more recent times, a growing appreciation of pollution hazards and the desire for a greater return on raw material costs has stimulated dairy industries to seriously consider new ways of using the large volumes of whey. Since there is a very limited market for lactose, attention has been directed to the hydrolysis of lactose to produce the more commercially valuable sugars, glucose and galactose. The hydrolysis can be obtained by much the same process as that described for the hydrolysis of milk lactose, i.e. using beta-galactosidase.

A number of pilot plants and industrial scale plants are in operation in various countries around the world, converting cheese whey into commercially viable products. In Britain, the Milk Marketing Board in collaboration with Corning Glass Works Inc., set up a semi-industrial plant in 1978, using beta-galactosidase from *Aspergillus niger* immobilized within porous silica beads of precisely controlled porosity. The immobilized enzyme was incorporated in a vertical cylindrical column, and operated continuously for five days per week,

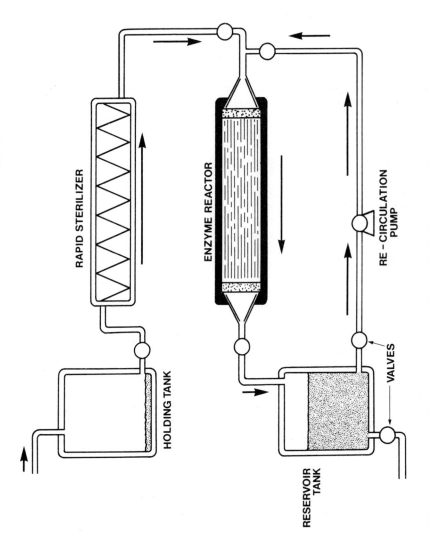

Fig. 5.4 A re-circulation reactor system for the hydrolysis of lactose in milk.

processing 10 000 litres of whey each day. As with the hydrolysis of milk lactose, the immobilized enzyme proved to be very stable, and the efficiency of catalysis was around 80% hydrolysis. The hydrolysed whey is normally concentrated to produce a protein-rich sweetener, which has about one third the sweetness of sucrose.

The unique combination of protein and sugar in the hydrolysed whey concentrate (HWC) provides distinct advantages, which increase the potential usefulness of the product above that of a simple sweetener. Protein is added to food products to improve emulsification, binding of constituents, foaming properties, gelling properties and control of texture (softness). The protein in the whey concentrate will fulfil these characteristics, and since it is also acid stable it can be used in acidic products. Whey concentrate has been used to reduce the content of traditional egg protein and milk solids (lactose) in a wide range of cakes, cake mixes, sponges and biscuits. In ice-cream and desserts such as mousse and yoghurt drinks, the protein imparts desirable characteristics such as 'mouth-feel' smoothness and 'body'. On a different tack, whey concentrate is a good supporting substrate for the growth of organisms such as Baker's yeast and bacteria, and may therefore be useful to replace or reduce the amount of traditional growth media in a range of industrial fermentations.

5.6   Production of amino acids

The last twenty years or so have seen a steady increase in the market for amino acids. They are used in research work, as additives in important fermentations, as food and animal feed supplements, and as ingredients in medicinal preparations. In addition, they have been incorporated into several cosmetic preparations such as facial cream/oil and hair shampoo products.

Amino acids can be produced by both chemical and biosynthetic procedures. The latter is the more expensive, but the former has the main disadvantage that it produces a racemix mixture of D and L stereoisomers of the amino acids. To obtain the L-isomer (the one that occurs in nature) it is necessary to resolve the mixture, and the most efficient method of resolution uses the selective stereo-specificity of enzymes. The underlying principle involved in the specific isolation of L-amino acids is quite simple. The mixture of D- and L-amino acids obtained by organic synthesis is chemically acetylated so that each amino acid molecule contains an acetyl group. The enzyme aminoacylase is then employed to specifically remove the acetyl groups from the L-amino acids, leaving the D-acetyl amino acids unchanged (Fig. 5.5). There is a fundamental solubility difference between acetylated amino acids and non-acetylated amino acids, so it is a fairly straight-forward chemical manipulation to isolate the L-amino acids.

Before 1969 the enzyme reaction was performed in a batch operation using soluble enzyme. Since 1969 producers have employed immobilized amino-acylase preparations in continuous operations. Several methods of immobilization have been evaluated, but the method used by the larger producers, particularly in Japan, is that of ionic binding to DEAE-Sephadex. This method of enzyme immobilization is simple and inexpensive. It has the added advan-

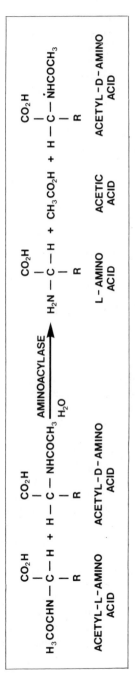

Fig. 5.5 The resolution of a racemic mixture of L- and D-amino acids using aminoacylase.

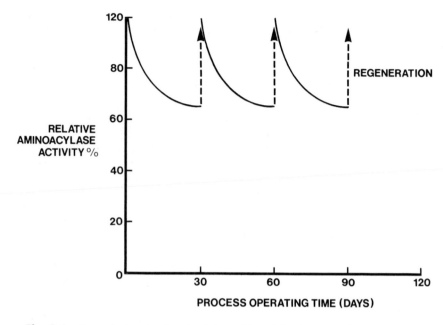

Fig. 5.6   A graph showing the steady loss of immobilized aminoacylase activity with time, and the regeneration of activity by addition of fresh soluble enzyme (from Chibata *et al.*, in Mosbach, 1976).

tage of simple regeneration of enzyme activity lost during the production. In continuous operation, 30% of the aminoacylase activity is lost from the reactor column after about 30 days use. Regeneration is accomplished by stopping the process for a short time while fresh soluble aminoacylase is added to the reactor under conditions to promote binding to the DEAE-Sephadex. The regeneration is performed every 30 days or so, and causes minimal disruption to the continuous production of amino acids (Fig. 5.6).

Probably the most important amino acid product in terms of bulk production is L-glutamic acid, as annual production now exceeds 0.5 million tonnes. In the form of monosodium L-glutamate it is used extensively world wide as seasoning in the food preparation industries, and as much as one gram per serving may be used in some food preparations 'to bring out the full natural flavour of the food'. L-glutamic acid is also used as a starting material in the synthesis of several speciality chemicals, for example N-acylglutamate is a useful biodegradable surfactant, used as a cleansing agent in various cosmetics, soaps and shampoo formulations because it exhibits a low degree of skin irritation. Another derivative of L-glutamic acid called oxopyrrolidine carboxylic acid is used as a moisturizer in cosmetic preparations because it acts in cooperation with other amino acid additives to promote water retention in the upper epidermis layer of skin (see Fig. 5.7). The development of L-aspartylphenyl-alanine (aspartame) as a sweetener has substantially increased the requirement for both L-phenylalanine and L-aspartic acid. Other amino acids have important uses as additives in food and animal feed products, pharmaceutical preparations and specialist medical products.

5.7 Cheese manufacture

The starting material for the manufacture of cheese is normally pasteurized whole milk. Other starting milks include non-fat skimmed milk for cottage cheese, and whole milk to which cream has been added for cream cheese. In addition to bovine milk, it is possible to make cheese from the milk of goat, sheep, buffalo, reindeer and camel. For some reason it is very difficult to make cheese from the milk of the ass, and apparently even more difficult to get the ass to donate the milk in the first place. There are only three categories of cheese (soft, hard-pressed and blue), although differences in the production technique, maturity, colour, shape etc., give rise to over 3000 varieties.

The process begins with the addition to the milk of a 'starter' culture of lactic acid bacterium (normally *Streptococcus lactis*). The milk is normally stirred for at least 30 minutes to allow the bacteria to convert the milk lactose to lactic acid, thus lowering the pH of the milk to pH 5.2–5.5. Next an enzyme complex called rennet is added, which acts on the milk proteins and causes the formation of a gel or coagulum that forms the solid curd. The curd is separated from the liquid whey, and the final stage is the ripening of the curd into an edible product.

Traditionally, rennet is a salt extract of the fourth stomach of calves, which are slaughtered before they are 32 days old. The principal enzyme in the extract is the acid protease chymosin (old name rennin). When a calf is about 4 weeks old it begins to eat roughages such as hay and grass, then chymosin is replaced by a more rigorous acid protease, pepsin. Therefore, depending on the age of the calf, it is possible to obtain rennet preparations varying in composition from 95% chymosin:5% pepsin to 5% chymosin:95% pepsin.

The major proteins in milk are the caseins, a group of small phosphoproteins that have a great capacity for association to form aggregates (called sub-micelles), containing about 35 casein molecules. In combination with calcium phosphate, 15 or more sub-micelles can associate to form a larger colloidal aggregate called the casein micelle. There is normally about one billion casein micelles per millilitre of milk, and these are responsible for characteristic 'milky' appearance. It is thought that the rennet protease attacks the casein protein on the surfaces of the sub-micelle, exposing 'sticky patches' on the micelle. The micelles aggregate together via the 'sticky patches' to produce the coagulum or gel. When the curd is collected it normally retains a small percentage of the rennet, which goes on to play a part in the ripening or curing of the curd.

Most of the characteristic flavour and texture of cheese is produced during the ripening stage. This process is dependent upon the enzyme-catalysed formation of a wide range of compounds from the main precursors found in the curd, i.e. protein, fats and sugar (Table 5.2). The enzymes involved in ripening

Table 5.2   Some enzyme catalysed transformations in cheese ripening.

			Effect on cheese	
Enzyme Source	Substrate	Products	Texture/Body	Flavour/Aroma
Rennet		Peptides	yes	–
Milk peptidases		Amino acids	yes	–
Microbial proteases	Proteins	Amines	yes	yes
Added decarboxylases		Ammonia	–	yes
Milk enzymes		Fatty acids	yes	yes
Microbial esterases	Fats	Esters	–	yes
Added lipases				
Milk enzymes	Sugars	Ketones	–	yes
Microbial enzymes		Aldehydes	–	yes
		Alcohols	–	yes

include chymosin, and many of the 20 or so indigenous milk enzymes, also enzymes present in the 'starter' bacteria. In some of the processes, additional enzymes (e.g. lipases) are added to produce characteristic flavours. The ripening process is controlled very precisely in order to determine the exact rate and level of enzyme activities during the period of the ripening. This ensures that the various enzymes do not 'over-react', producing bitter-tasting compounds, off-smelling odours or poor texture.

The principal development in the enzyme technology associated with the manufacture of cheese has been concerned with replacing the expensive and limited supplies of rennet. A wide range of substitutes have been evaluated, including bacterial proteases, plant proteases, bovine pepsin and pig pepsin. Only a few have proved useful, in particular the proteases from some *Mucor* species, which have similar catalytic properties to that of chymosin and are used in modern cheese manufacture.

5.8   Tenderization of meat

In commercial meat processing it is desirable to transfer carcasses into cold storage, as this reduces weight loss by evaporation, and retards microbial growth. However, storage leads to the development of toughness in the carcass due to biochemical changes in the muscle tissues. After slaughter, anaerobic glycolysis continues to operate in the muscle, converting glucose to lactic acid, and lowering the pH of the muscle. In beef the pH drops slowly during storage, but when it reaches pH 5.5, the physiological integrity of the muscle is lost. The muscle proteins become locked in a contracted state, stiffening the muscle to produce a state of rigor mortis. For further information on the biochemistry and physiological changes associated with rigor mortis see Winstanley (1986). This and other effects cause toughening of the muscle tissue, and rigor mortis is normally set in beef about 18–24 hours after slaughter.

Regain of tenderness can be accomplished in beef by continuing the storage for 10–12 days to allow onset of the natural process of autolysis involving cellular lysosomes. This process (called conditioning) involves a group of lysosomal proteases called cathepsins, which break down the connective tissue and myofibrillar proteins to weaken the mechanical structure of the meat. The conditioning must be undertaken under controlled bacteriostatic conditions – this requirement and the time involved, increases the cost of meat processing.

The addition of proteases to influence the tenderization of meat is a long established practice. By far the most widely used enzyme is papain from the latex of the *Carica papaya* plant. Papain can be added prior to slaughter by injection or after slaughter. Post-slaughter addition allows greater control over the degree of tenderization achieved. The amount of time in storage is reduced, because an appreciable amount of tenderization is accomplished during cooking before the enzyme is inactivated at around 90°C. The general trend in the meat industry is toward more rapid processing (e.g. hot boning), and it is likely that the time delay required for natural conditioning will be circumvented by an increase in the application of enzyme tenderizers. The increase in home freezer storage of large meat cuts has established a market in meat tenderizers (papain) for home application.

5.9   Leather industry

Protease enzymes are used in the processing of skins and hides into leather, in particular for the removal of hair and wool, and for increasing the pliability (called bating) of leather. The traditional chemical procedure for hair removal from cow hide employs calcium hydroxide (lime) and sodium sulphide. The lime is used to swell and soften the skin, and the sodium sulphide is used to disrupt and weaken the hair follicle so that the hair may be readily washed off the skin. This method is cheap and effective, but unsatisfactory because it generates toxic hydrogen sulphide (H_2S) gas during the process. The sulphide can be replaced by an alkali-stable protease enzyme from *Bacillus* species. The enzyme catalyses the breakdown of the protein keratin in the hair, allowing the hair to be easily removed (Fig. 5.7).

The process for sheepskin is rather more gentle in order to avoid damage to top quality wool that can be spun into valuable wool yarn. Liquid protease preparations such as pancreatic extracts of trypsin and chymotrypsin are applied to the flesh side of the sheepskin. The proteases diffuse to the basal papilla of the hair follicle, and subsequent protein breakdown releases the wool hair with minimal damage (Fig. 5.7).

Bating of leather to make it more pliable and workable was, until 70 years ago, accomplished using cow or horse dung, which contains several pancreatic-derived proteases such as trypsin and chymotrypsin. Such work was unpleasant and hazardous with a high risk of infection. A simple appreciation of the origin of the enzymes in the dung led to the present practice of extracting the enzymes directly from the pancreas of slaughtered cattle and pigs. The extent of bating depends on the time allowed for the proteases to act on the skin proteins. This is

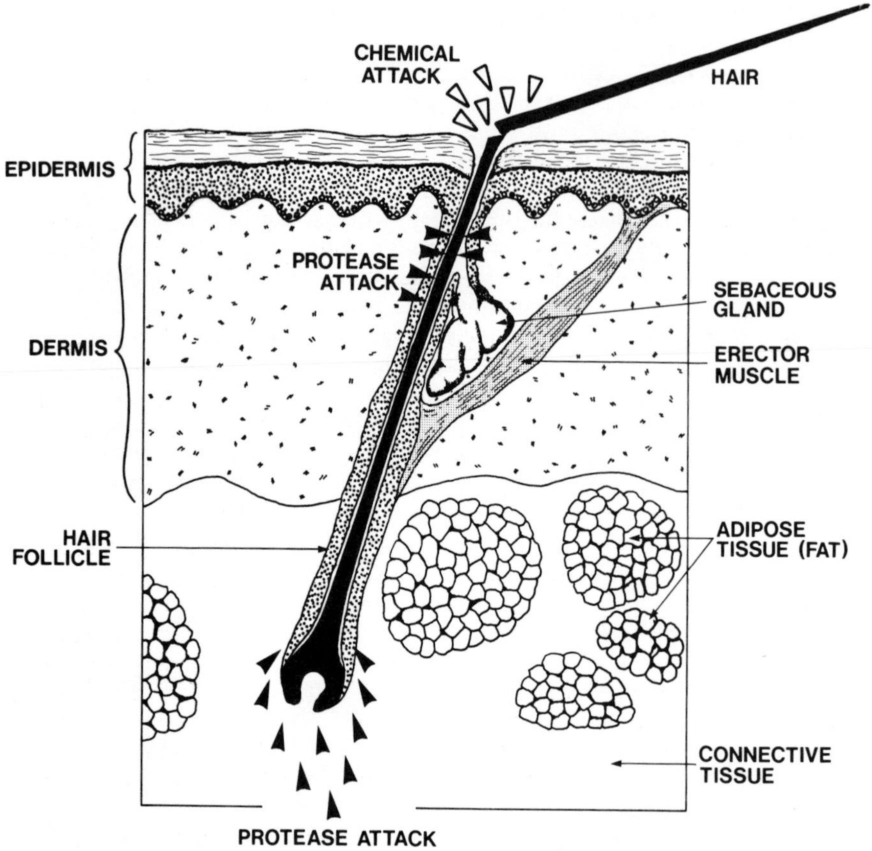

Fig. 5.7 A diagram showing the use of chemicals and the use of proteases to remove hair from a skin.

dictated by the amount of flexibility required in the finished product. For example, glove leather requires more bating than handbag leather, which in turn requires more bating than shoe leather. The actual mechanism involved in the softening of leather has not been fully elucidated, but clearly involves some partial disintegration of the structural protein (collagen) of connective tissue in the skin.

5.10   Fruit juice and wine production

A large spectrum of berries and fruits are used in the production of a wide range of fruit juice products. Further, a substantial extension of the range is provided by the incorporation of an additional (fermentation) process to produce alcohol, and transform the fruit juice into wine. Enzymes are used to increase the yield of fruit juice during extraction, and to produce the required degree of clarity for the final product.

Extraction normally involves mashing or grinding the fruit, and during this process enzymes are added to degrade the pectin complexes that tend to reduce the yield of juice from a fruit. Pectin is a complex mixture of at least three polysaccharide components. The principal polysaccharide is polygalacturonic acid and this is normally combined with polygalactose and the highly branched polyarabinose. In addition, the carboxyl side groups of galacturonic acid are partly esterified with methanol residues. Pectins are found within and between the cell walls of higher plants, where they cement the cellulose fibres together to form a rigid structure. In unripe fruit, pectins remain insoluble and the juice is free to run. However, as part of the natural ripening and softening of the fruit, pectin is partly degraded and solubilized by indigenous pectinases. The soluble portion of the pectins extends into the juice while the insoluble end remains attached to the cellulose. The presence of this polysaccharide increases the viscosity of the juice, making it less free to run, and difficult to separate from the flesh of the fruit (Fig. 5.8). The addition of pectinases and cellulases during the extraction breaks the link between the pectin and the cellulose, releasing the pectin into the juice where it is further degraded by pectinases to increase the fluidity of the juice. The extent of the problem varies depending on the fruit, for example pear and apple require limited application of enzymes, but blackcurrant and cranberries require greater use of enzymes to increase the yield of juice. There are a wide range of pectinase enzymes produced by fungi, plants and some bacteria. However the most commonly used enzymes are pectin esterase which demethylates the galacturonic acid residues, and polygalacturonase which hydrolyses the polysaccharide chains to promote depolymerization of pectin complex.

The pectin content of a juice is a principal factor in the formation of stable colloidal suspensions together with protein, starch and hemicellulose. For some products such as orange juice and pineapple juice, the cloudiness is a characteristic of the product and is therefore retained. However, in other products such as apple juice, pear juice and wine, it is necessary to clarify the product. Clarification can be achieved in a two-stage process designed to collapse and coagulate the colloids. Added pectinases further hydrolyse the pectin chains, increasing the viscosity, and destabilizing the suspensions. Addition of fining agents promotes the collapse and sedimentation of the suspensions. Pectinase enzymes from *Aspergillus niger* (which causes soft rot in fruits) are used in industrial processing, and in the production of home-made wine. The popularity of making home wine has created a significant market in the application of pectinases to improve juice extraction when starting from whole fruit, and to aid the clarification of the final product. With regard to the latter application, it is useful to remember that the juice contains natural pectinases, and that these enzymes are heat sensitive. Therefore, when preparing the ingredients for home wine fermentation, it is important to allow boiled water to cool down to 20°C before mixing with the fruit for extraction, or mixing with the juice prior to the fermentation. This will avoid heat denaturation of the

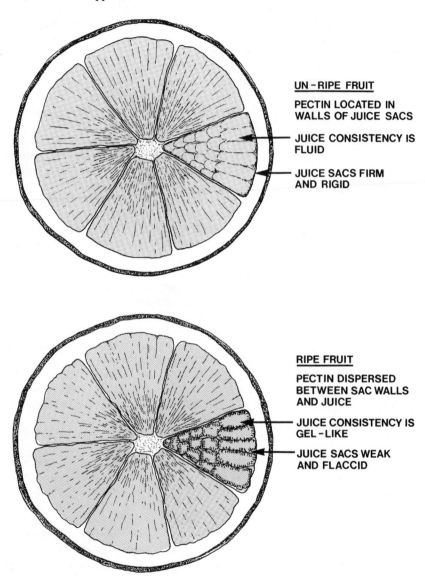

Fig. 5.8 Diagram showing a change in the structure of juice sacs in ripe fruit as a result of natural pectin degradation.

pectinase enzymes and enable pectin degradation to occur during the fermentation stage.

Another useful enzyme application in fruit juice processing is the hydrolysis of bitter-tasting compounds. Bitterness in the juice extracts is associated with a flavonone compound called naringin, which is very water soluble and is located in the juice sacs of the fruit (Fig. 5.8). The hydrolytic enzyme naringinase from *Aspergillus* is used to reduce the level of naringin in several citrus fruit juice products, particularly grapefruit juice. It is important to monitor and control the degree of de-bittering to ensure that bitterness is reduced to an acceptable level but not completely removed. Naringin is of some value because it stimulates taste buds in the mouth, producing a greater perception of taste so that food and beverages consumed after grapefruit require less addition of sugar to provide a sweet taste. A number of slimming diets include grapefruit as a starter in the hope that sugar intake might be reduced.

Hydrolysis of naringin is normally accomplished by addition of soluble naringinase during batch processing. However, after the required degree of hydrolysis it is necessary to inactivate the enzyme by heating at temperatures in excess of 90°C, and this step may affect the product adversely. Immobilized preparations of naringinase have been examined and found to be effective for hydrolysing naringin, and the process may be operated on a continuous system rather than a batch system. Also the heat inactivation step can be omitted because the immobilized enzyme is easily separated from the juice when the desired degree of de-bittering has been achieved.

A potential source of spoilage in commercial liquid products such as wine, beer, fruit juices and carbonated soft drinks is the inherent content of dissolved oxygen in the liquid. In the presence of oxygen, various coloured compounds and flavour-bearing substances in the product may be oxidized giving colour-fading and off-taste. In the case of bottled products, it is known that exposure to sunlight promotes the formation of H_2O_2 from dissolved oxygen, and that H_2O_2 readily oxidizes a wide range of compounds. In the case of tinned products, dissolved oxygen can promote oxidation of the metal (rusting), giving a metallic taste to the product. An effective solution to the problem is the addition of an antioxidant to the product, and in many liquid products the preferred antioxidant is an enzyme preparation containing glucose oxidase and catalase. The net result of the action of these two enzymes is that oxygen is used up in the conversion of glucose to gluconic acid as illustrated in Chapter 3 (see Fig. 3.6).

5.11 Baking industry

Flour is the principal ingredient in bread and bakery products, and the most common source of flour for baking is wheat grain. Wheat flour contains mostly starch with only small amounts of simple sugar such as glucose, some protein (gluten) and several enzymes. Basically, baking involves mixing flour with warm water, yeast and fat to make a dough, during which enzymes will degrade the starch to glucose. The mixture is incubated at a warm temperature to continue the starch hydrolysis, and allow the yeast to ferment the glucose to alcohol and CO_2, causing the dough to rise. Lastly the product is baked in a hot oven to heat inactivate the yeast and enzymes, and prevent further activity (Fig. 5.9).

The extent of the starch hydrolysis is vital, and depends on the level of natural amylases present in the flour. A low level of amylases produces a limited

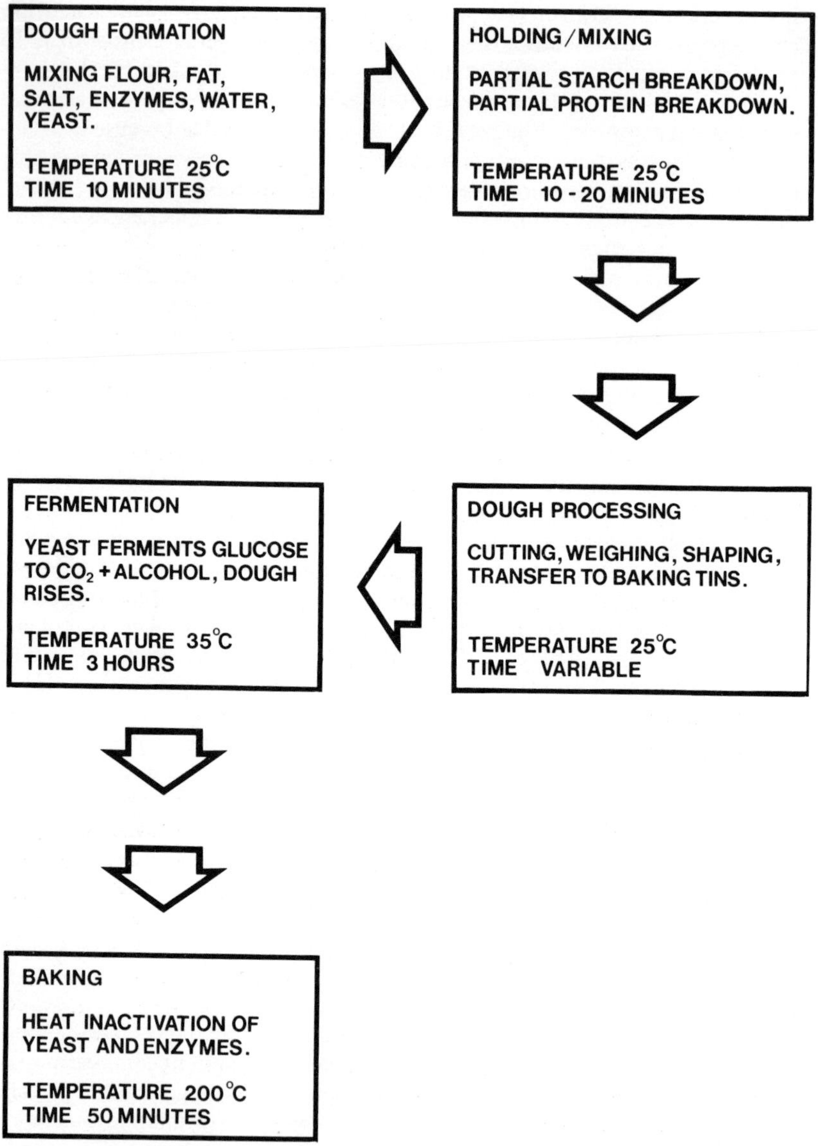

DOUGH FORMATION

MIXING FLOUR, FAT,
SALT, ENZYMES, WATER,
YEAST.

TEMPERATURE 25°C
TIME 10 MINUTES

HOLDING/MIXING

PARTIAL STARCH BREAKDOWN,
PARTIAL PROTEIN BREAKDOWN.

TEMPERATURE 25°C
TIME 10 - 20 MINUTES

FERMENTATION

YEAST FERMENTS GLUCOSE
TO CO_2 + ALCOHOL, DOUGH
RISES.

TEMPERATURE 35°C
TIME 3 HOURS

DOUGH PROCESSING

CUTTING, WEIGHING, SHAPING,
TRANSFER TO BAKING TINS.

TEMPERATURE 25°C
TIME VARIABLE

BAKING

HEAT INACTIVATION OF
YEAST AND ENZYMES.

TEMPERATURE 200°C
TIME 50 MINUTES

Fig. 5.9 A scheme of the principal stages in the process of baking.

release of glucose from starch, and leads to poor CO_2 formation by the yeast. The level of amylases in wheat flour depends on the growth conditions of the wheat, and storage conditions of grain. Generally flour is sufficient in beta-amylase (an exo-attacking enzyme) but it is usually necessary to add a fungal amylase (an endo-attacking enzyme) to the dough mixture to ensure extensive starch hydrolysis and release of glucose. A fungal alpha-amylase is used because it is readily heat inactivated at 60°C.

Enzymes are also employed in changing the physical viscoelastic properties of the dough to improve handling and machining, and enable a wider range of bakery products to be produced. The protein content of flour is largely responsible for the viscoelastic properties of the dough, and fungal proteases are added to partially hydrolyse the protein, thus altering the viscoelastic properties. Some wheat varieties are high in protein content and give a 'hard' flour, which is strong with limited elasticity. In the case of machine-produced cakes and biscuits, it is necessary to have a low protein 'soft' flour with good elasticity and suitability for machine handling. Fungal proteases are normally added to both 'soft' and 'hard' flours, because in addition to improving the viscoelastic properties, the proteolysis releases amino acids, which support the growth of yeast and enhance flavour in the product.

In the production of white bread the wheat flour is supplemented with a soy flour, which contains lipo-oxygenase enzymes. These enzymes use oxygen to convert polyunsaturated fatty acids such as linoleic acid and linolenic acid to hydroperoxide derivatives. These derivatives bleach the natural coloured pigments of the dough by oxidation reactions. In addition to losing some fatty acids, other constituents including some vitamins are destroyed by the oxidation so reducing the nutritive value of the product. For further details on the use of enzymes in baking and other food industries see Birch *et al.* (1981).

5.12   Brewing industry

As far as enzyme applications are concerned, the brewing industry has very much in common with the baking industry. In brewing the raw material is normally barley grain, but the aim (as in the baking industry) is to hydrolyse the starch to glucose for use by yeast in fermentation. Traditionally brewers allow the barley grains to germinate thereby producing the enzymes necessary for degradation of starch and of protein. Generally malted (germinated) barley produces sufficient levels of all the enzymes necessary for effective natural degradation of the starch without need of enzyme supplements. However, the natural process of germination is subject to biological variation, sometimes resulting in minor or major enzyme deficiency in the malt. It is always necessary to evaluate the enzyme proficiency of the malt, and supplement with added amylases and/or proteases if necessary.

The present trend in the industry is in favour of greater application of enzymes to offset the problems associated with the traditional preparation of malt. In particular, malting takes 8–10 days with no guarantee of a full complement of all the necessary enzymes, and the barley embryo consumes a proportion of the starch for growth so reducing the potential amount of glucose for alcohol production. Brewers have found that by reducing the period of germination to 4 days, and supplementing the malt with fungal amylases and proteases, it is possible to obtain a consistently reliable malt without altering the flavour of the end product.

Proteolytic enzymes are also used during beer finishing operations as chill-proofing agents to ensure the long-term brilliance and colloidal stability of the final product. Chill-proofing refers to the treatment of beer to avoid the formation of chill-haze during cold storage. The haze is due to the cold precipitation of protein/polyphenol and protein/tannin complexes, giving a dull cloudy appearance to the beer. Proteases prevent haze formation by partially degrading the protein necessary for haze formation. The plant enzyme papain is the most widely employed chill-proofing protease because it catalyses only a limited breakdown of the protein. A disadvantage in the use of protease for the chill-proofing stage is that low levels of enzyme activity may remain after pasteurization, which is supposed to inactivate the enzyme. During long periods of storage, protein breakdown may continue and generate a sharp bitter taste due to increased amounts of amino acids. Also the desirable foaming of beer that occurs during pouring to form a 'head' on top of the beer is due to the protein component in the beer. Thus excessive protein breakdown is undesirable because it results in poor 'head' formation/retention in the beer. Aficionados of beer drinking accurately describe such beer as typically 'a bit flat'.

5.13 Recombinant DNA technology

This is a powerful technique that has proved very useful for investigating the detailed structure, function and regulation of DNA molecules (see Warr, 1984). An important spin-off from this work is the identification, isolation and manipulation of specific eukaryotic genes for application in industry and medicine. In particular, the technique provides a means of transferring functional genes from one organism to another. In medicine, it is envisaged that this approach could eventually lead to the development of a gene replacement therapy for human genetic disorders. In industry, the possible applications are more readily attainable in the short term, as indicated by the successful cloning of eukaryotic genes in bacterial cells to produce important proteins. The eukaryotic gene contains the instructions for making the protein, and is inserted into bacterial cells. Bacteria grow and multiply rapidly thus producing multiple copies of the eukaryotic gene, which are then translated by the bacterial protein synthesizing machinery to produce the desired protein product. For example, consider the production of insulin by this approach. First, the functional eukaryotic gene that codes for insulin is identified and cut out of the eukaryotic chromosome. Then a DNA hybrid molecule is constructed consisting of the insulin gene and bacterial DNA. Direct insertion into a bacterial chromosome would cause complications for the bacterium, so the insulin gene is inserted into the DNA of a vector such as a plasmid. A plasmid is a small circular extra-chromosomal DNA molecule found in the cytoplasm of many bacterial cells, and it is capable of independent self-replication. Lastly, the plasmid containing the insulin gene is returned to the bacterial cell, and the bacteria cultured in bulk to generate many copies of the insulin gene, which in turn lead to the production of large amounts of insulin.

The approach is absolutely dependent on the use of enzyme tools for cutting out genes from DNA molecules, and patching the excised gene into the vector DNA to construct the recombinant DNA molecule. The DNA cutting enzymes are called restriction endonucleases, and they cut both strands of double helix DNA by hydrolysing the phosphodiester bonds between nucleotides. They are found very widely in prokaryotic cells, and the biological function of such enzymes is to protect prokaryotes such as bacteria from invading viral (bacteriophage) DNA. Native bacterial and plasmid DNA is extensively methylated, which serves to protect the bacterial DNA from the restriction enzymes. More than 200 different enzymes have been isolated from a wide range of prokaryotic cells, and about 120 of these are commercially available. A nomenclature for naming the enzymes is in use, and involves a three letter abbreviation of the source, i.e. the bacterium, followed by a designation of the strain (if appropriate), and a Roman numeral to indicate the existence of more than one enzyme from a given source. Thus, restriction endonuclease *Eco* RI is isolated from *Escherichia coli* strain RY 13, and it is enzyme I from this organism.

Restriction enzymes are highly specific and are able to recognize short sequences of DNA of only four to six base pairs in length. The enzymes cleave the DNA at these particular recognition sequences, and over 70 different recognition sequences have been identified. Since a particular sequence may occur many times in a DNA molecule, an enzyme may chop DNA into a number of restriction fragments. Therefore, by using a combination of enzymes it is possible to obtain a fragment of DNA consisting of an intact functional gene. In most cases the sequence specificity is dictated by a palindrome mirror image structure. A palindrome is a sequence of bases that reads the same left to right as it does right to left. For example, the specific sequence that is recognized by the *Eco* RI restriction enzyme is a six nucleotide palindrome mirror image structure:

> 5'-G-A-A-T-T-C-3'
>
> **: : : : : :**   DNA double helix
>
> 3'-C-T-T-A-A-G-5'

Restriction enzymes cleave both chains of a DNA double helix, but the position of cleavage on each chain can be the same or it can be different. If the cleavage is symmetrical then flush-ended DNA fragments are produced, and if the cleavage is not symmetrical then staggered or cohesive ends are produced (Fig. 5.10). The cohesive or 'sticky ends' may have 3, 4 or 5 base single-strand DNA tails, which are very useful for enabling the insertion of a DNA fragment into the plasmid DNA. Thus, if the same restriction enzyme that was used to obtain the sticky ends on the DNA fragment is also used to open-up the plasmid DNA, then, because of the enzyme specificity, the sticky ends on the DNA fragment will match the sticky ends produced in the plasmid DNA, so enabling the two to be joined forming a recombinant DNA molecule. Several enzymes that recognize the same palindrome sequence are called isoschisomers, and Fig. 5.10 shows *Xma* I from *Xanthomonas malvacearum* and *Sma* I from

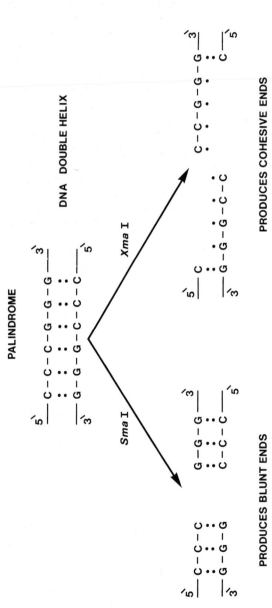

Fig. 5.10 A diagram showing two restriction enzymes that recognize the same palindrome sequence of DNA, but cleave the DNA differently.

Serratia marcescens producing cohesive ends and flush ends respectively from the same palindrome sequence of DNA. Another enzyme DNA ligase, which is normally involved in the replication of DNA, is used to join the ends of the DNA molecules together.

From an industrial view, the approach has a significant potential for producing important proteins that are normally very expensive because in cells and tissues they occur naturally in very small quantities. In addition to the protein hormones such as insulin, growth hormone and thyroid stimulating hormone, other key proteins that may be produced include anti-viral agent interferon and blood clotting Factor VIII. The production of human enzymes would be of considerable use in enzyme replacement therapy, as well as making available catalytically useful enzymes that are not used because they are too expensive to extract from cells or tissues. Also since single plant cells are capable of regenerating to form an intact plant, it is possible to envisage a new era of plant breeding at the molecular level by manipulation of plant genes to produce better varieties of plant. It should be possible to increase plant resistance to pathogens by inserting specific genes responsible for resistance. It may be possible to extend the capacity to fix atmospheric nitrogen to cereal and other important crops, and introduce cold tolerance to tropical or sub-tropical plants, thereby creating considerable improvements in crop utilization and performance.

There are clearly a great many possibilities for new developments in a whole range of industries, and this will create a new era for advancement of biological science (see Smith, 1981).

References and Further Reading

Birch, G.G., Blakebrough, N. and Parker, K.J. (eds) (1981). *Enzymes and Food Processing*. Applied Science Publishers, London.

Boswell D.R. and Bathurst I.C. (1985). Molecular physiology and pathology of alpha$_1$-antitrypsin. *Biochemical Education*, **13(3)**, 98–104.

Brown, M.S. and Goldstein, J.L. (1984). How LDL receptors influence cholesterol and atherosclerosis. *Scientific American*, **251(5)**, 52–60.

Brown, M.S. and Goldstein, J.L. (1986). A receptor-mediated pathway for cholesterol homeostasis. *Science*, **232**, 34–47.

Carr, P.W. and Bowers, L.D. (1980). *Immobilized Enzymes in Analytical and Clinical Chemistry*. John Wiley and Sons, New York and London.

Chang, T.M.S. (ed.) (1977). *Biomedical Applications of Immobilized Enzymes and Proteins*, vols 1 and 2. Plenum Press, New York and London.

Godfrey, T. and Reichelt, J. (eds) (1983). *Industrial Enzymology*. Macmillan, London.

Goldstein G.W. and Betz A.L. (1986). The blood-brain barrier. *Scientific American*, **225(3)**, 70–9.

Gronow, M. and Bliem, R. (1983). Production of human plasminogen activators by cell culture. *Trends in Biotechnology*, **1**, 26–9.

Hartley, B.S., Atkinson, T. and Lilly, M.D. (eds) (1983). *Industrial and Diagnostic Enzymes*. The Royal Society, London.

Hough, L. and Emsley, J. (1986). The shape of sweeteners to come. *New Scientist*, 19 June 1986, 48–54.

MacKensie, D. (1984). Leprosy: the beginning of the end. *New Scientist*, 3 May 1984, 30–3.

Mosbach, K. (ed.) (1976). *Methods in Enzymology*, vol. 44. Academic Press, New York and London.

Renneberg, R., Schubert, F. and Scheller, F. (1986). Coupled enzyme reactions for novel biosensors. *Trends in Biochemical Sciences*, **11**, 216–20.

Russell, G.E. (ed.) (1984). *Biotechnology and Genetic Engineering Reviews*, vol. 1. Intercept, Newcastle upon Tyne.

Sibbald, A. (1983). Chemical-sensitive field effect transistors. *I.E.E. Proceedings*, **130**, part 1 (5), 233–44.

Smith, J.E. (1981). *Biotechnology*. Studies in Biology, no. 136. Edward Arnold, London.

Trevan, M.D. (1980). *Immobilized Enzymes*. John Wiley and Sons, New York and London.

Warr, R.J. (1984). *Genetic Engineering in Higher Organisms*. Studies in Biology, no. 162. Edward Arnold, London.

Wilson, K. and Goulding, K.H. (eds) (1986). *Principles and Techniques of Practical Biochemistry*. Edward Arnold, London.

Winstanley, M. (1986). The tender touch of rigor mortis. *New Scientist*, 29 May 1986, 36–9.

Woodward, J. (ed.) (1985). *Immobilized Cells and Enzymes*. IRL Press.

Wynn, C.H. (1979). *The Structure and Function of Enzymes*, 2nd edition. Studies in Biology, no. 42. Edward Arnold, London.

Index